QUESTIONING
THE MILLENNIUM

Stephen Jay Gould is the Alexander Agassiz Professor of
Zoology and Professor of Geology at Harvard University,
and the Curator for Invertebrate Palaeontology in the
University's Museum of Comparative Zoology. His
publications include *Eight Little Piggies*, *Wonderful Life*,
Ever Since Darwin, *The Panda's Thumb*, *Dinosaur in a
Haystack* and, most recently, *Life's Grandeur*.

Detail from *The Last Judgement* (1536–1541), Michelangelo.

Stephen Jay Gould

QUESTIONING THE MILLENNIUM

A Rationalist's Guide to a Precisely Arbitrary Countdown

VINTAGE

Published by Vintage 1998

2 4 6 8 10 9 7 5 3 1

Copyright © Stephen Jay Gould 1997

The right of Stephen Jay Gould to be identified as the author of this work has been asserted by him in accordance with the Copyright, Designs and Patents Act, 1988

Part 2 was originally published, in different form, in *Dinosaur in a Haystack* by Stephen Jay Gould. Copyright © 1995 by Stephen Jay Gould

First published in Great Britain by
Jonathan Cape in 1997

Vintage
Random House, 20 Vauxhall Bridge Road,
London SW1V 2SA

Random House Australia (Pty) Limited
20 Alfred Street, Milsons Point, Sydney
New South Wales 2061, Australia

Random House New Zealand Limited
18 Poland Road, Glenfield, Auckland 10,
New Zealand

Random House South Africa (Pty) Limited
Endulini, 5A Jubilee Road, Parktown 2193,
South Africa

Random House UK Limited Reg. No. 954009

A CIP catalogue record for this book
is available from the British Library

ISBN 0 09 976581 0

Papers used by Random House UK Ltd are natural, recyclable products made from wood grown in sustainable forests. The manufacturing processes conform to the environmental regulations of the country of origin

Printed and bound in Great Britain by
Mackays of Chatham PLC, Chatham, Kent

In loving memory of my friend Carl Sagan

The most passionate rationalist of our times
The best advocate for science in our millennium

CONTENTS

QUESTIONING THE MILLENNIUM

PREFACE

OUR PRECISELY ARBITRARY
MILLENNIUM

I began to think about this book during the first week
of January 1950. I was eight years old, and a good
part of my life revolved around the simple pleasures of
weekly rituals. On Sundays, I would pull out *The New
York Times* sports section and turn to the agate-type list-
ings of performance records for major league baseball
players. I would take an index card, align all the stats
for a single player along the top edge, and then slowly
move the card down, a player at a time, studying the
numerical data for each in turn.

The weekly arrival of *Life* magazine, that quintessen-
tial organ of middlebrow culture, defined a second
activity—this time a less structured survey of pictures.
The first issue for 1950 hit me with a force that I still
don't comprehend, and burned into my cortex a permanent memory as potent and enduring as the records
of childhood's more tumultuous events—my kid
brother's birth, my father's return from war. This first

issue for 1950 marked the halfway point of the twentieth century by evaluating what had happened and predicting what the second segment might bring. (The publication of this special issue in January 1950, rather than January 1951—the "true" half-century point, according to one school of thought—provides yet another expression of that recurring, perverse, frustrating, funny, yet somehow fascinating debate on the unresolvable issue of when centuries end, the subject of Part 2 in this book and the source of more passionate discussion than ever before, because the forthcoming passage also marks the inception of a new millennium.)

For some reason, as I scanned this issue, my main thought went forward to the year 2000. My third grade mathematics told me that I would then be fifty-eight years old, while two living grandparents testified to the high probability that I would witness this far more interesting event. I have been buoyed by this lovely idea ever since—that I would enjoy the rare privilege of experiencing a transition that (however arbitrary) would rivet the attention of nearly all nations. Most folks live and die in years of little numerical distinction. I figured that I was one helluva lucky guy. When I should have died of cancer in the mid-1980s, but recovered instead, I listed only two items as placeholders of all the reasons for cherishing life in our times: "I dwelled on many things—that I simply had to see my children grow up, that it would be perverse to come this close to the millennium and then blow it" (from the preface to *The Flamingo's Smile,* 1985).

There will be an orgy of millennial books, and I hate to follow crowds. What then, beyond the indulgence of a little boy's whim dating from January 1950, can possibly justify my addition to this ephemeral genre? In one sense, this little book rests its case for distinctiveness upon an omission. I will eschew, absolutely and on principle, the two staples of fin de siècle literature, especially of the apocalyptic sort inspired by a millennial transition. I regard these subjects as speculative, boring, and basically silly—for they rank as primary examples of "punditry's" fundamental error: the fatuous notion that a head-on rush at the biggest questions will automatically yield the deepest insights.

I shall, first of all, make *no predictions* about human futures, either for years, decades, millennia, or geological ages; or for individuals, family lineages, or races; or for cities, nations, hemispheres, or galaxies. (I limit myself to predicting the aforementioned glut of books about the millennium.) Second, I refuse to speculate about the psychological source either for the angst that always accompanies the endings of centuries (not to mention millennia) or for the apocalyptic beliefs that have pervaded human cultures throughout recorded history, particularly among the miserable and malcontented.

Instead, I will confine myself to a set of related millennial questions that may seem paltry or laughably limited compared with the grandeur of unknowable futures, but that (as I hope to convince you) gain greater potential import by their definability and their exemplification, in fruitful ways, of questions as general as the nature of truth and the mechanisms of human knowledge. God bless all the precious little examples and all their cascading implications; without these gems, these tiny acorns bearing the blueprints of oak trees, essayists would be out of business. I want to talk about calendars and numbers; about fingers, toes and the perception of "evenness"; about the sun and the moon, the age of the earth, and the birth of Jesus.

These preciously definite, but wondrously broad, calendrical questions all arise from a foible of human reasoning, and also underlie all the passionate arguments now swirling around the impending millennial transition. In a famous motto, the Roman dramatist Terence stated in the second century B.C.: *"Homo sum: humani nihil a me alienum puto"* (I am a man, and nothing human can therefore be alien to me). Our urge to know is so great, but our common errors cut so deep. You just gotta love us—and you gotta view misguided millennial passion as a primary example of our uniqueness and our absurdity—in other words, of our humanity.

The astronomical, historical, and calendrical questions of this book all rest upon the distinction between nature's factual status and our arbitrary definitions within these constraints—in other words, the interaction of undeniable reality and the flexibility of human interpretation. Some things in nature just are—even though we can parse and interpret such real items in wildly various ways. A lion is a lion is a lion—and lions are more closely tied by genealogy to tigers than to earthworms. (Of course, I recognize that some system of human thought might base its central principle upon a spiritual or metaphorical tie between lion and earthworm—but nature's genealogies would not be changed thereby, even though the evolutionary tree of life might be utterly ignored or actively denied.)

But other important categories in our lives, however precisely definable and however objectively ascertainable, must be judged as arbitrary in the crucial sense that nature permits a plethora of equally reasonable alternatives, while providing no factual basis for a preferred choice. For example, each pitched baseball crosses home plate in a particular location of undeniable factuality—but the definitions for balls and strikes are human decisions, entirely arbitrary with respect to the physics of projection, however sensible within a system of rules and customs regulating this popular sport. (These definitions can also change—and have often done so—when circumstances favor an alteration.) Similarly, although nature dictates days by a full rotation of the earth, the parsing of days into packages of seven, called weeks, represents an arbitrary decision of some human cultures.

Millennial questions record our foibles, rather than nature's dictates, because they all lie at the arbitrary end of this spectrum. At the opposite and factual end, nature gives us three primary cycles—days as earthly rotations, lunations (we define our months slightly differently, and for interesting reasons) as revolutions of the moon around the earth, and years as revolutions of the earth around the sun. (God—who, on this issue, is either ineffable, mathematically incompetent, or just plain comical—also arranged these primary cycles in such a way that not a one of them works as a simple multiple of any other—the major theme of Part 3 and a source of many millennial issues.)

In an intermediary position, definitions are surely arbitrary, but nature's factuality nudges independent cultures toward common (but by no means universal) resolutions. The solar year, for example, does not fall naturally into four equal periods called seasons, but the existence of two solstices and two equinoxes—ascertainable with reasonable ease in most places where people live at high density, and truly important to know for such basic activities as hunting and gathering, and the later development of agriculture—may impose a slight natural bias for division by four.

Nonetheless, many cultures use other systems more attuned to immediate surroundings. In many tropical regions, for example, day lengths and temperatures don't vary drastically, and solstices and equinoxes may regulate nothing of great importance—whereas a two (or more) fold division of predictable rainy and dry times within the solar year makes far more sense as a basis for divisions. I once spent several months on Curaçao, the formerly Dutch island off the coast of Venezuela. Here no prominent seasonality exists in any natural form (though an indirect surrogate might be found in fluctuating numbers of tourists from lands with pronounced climatic cycles), for the trade winds blow all year from the east, and dryness always prevails. The daily newspaper doesn't even include a weather report, for nothing much varies. Any notable fluctuation—a hurricane, or even an extensive storm—is treated as news, not weather.

Millennial madness (or at least fascination) surely

lies at the arbitrary end of this spectrum, for nature recognizes no divisions by thousands. The intrinsic advantages of decimal mathematics have often been noted,
and our Arabic numerology surely gives 1,000 that nice
look of evenness (enhanced in our century by the active
turning of automobile odometers). But we also recognize that these advantages do not arise from nature's
construction, and we know that several cultures developed entirely functional (and beautifully complex)
mathematical systems on bases other than 10—and,
therefore, with no special status attached to the number 1,000 at all.

Perhaps the old saw that links decimal mathematics
to our ten fingers has validity after all, and perhaps, for
this reason, systems based on ten do follow a natural
bias. But Mayan culture, for example, developed an elegant vigesimal mathematics based on 20—perhaps they
counted both fingers and toes!—and this complex
numerical system honored many cycles and "even-
nesses," but not millennia or any multiples of 1,000.
Besides, and in any case, our ten fingers represent an
evolutionary contingency that might easily have settled
upon a different and equally functional outcome. Darwinian processes did not confer ten fingers upon early
reptiles because, more than 300 million years later, a
brainy species would walk upright, separate fingers

from toes, and then recognize that ten fingers imply the most convenient mathematics! The first terrestrial vertebrates had six, seven, or eight digits on each limb—the *Eight Little Piggies* of one of my previous books. Base 8 isn't bad either—but vertebrates followed a different evolutionary pathway.

And maybe, on a plausible alternative earth, the horse would not have become extinct in North America. The Mayans might then have domesticated a beast of burden, invented the wheel, and maybe even those two great and dubious innovations of ultimate domination—efficient oceanic navigation and gunpowder. Europe was a backwater during the great Mayan age in the midst of the first millennium of our Christian era. Continue the reverie, and Mesoamerica moves east to conquer the Old World, makes a concordat with Imperial China—and vigesimal mathematics rules human civilization for the foreseeable everafter. The millennium—the blessed thousand-year reign of a local god known as Jesus Christ—then becomes a curious myth of a primitive and conquered culture, something that kids learn in their third grade unit on global diversity.

But decimal Europe prevailed instead. And decimal Europe became Christian for other contingent reasons. And Christianity has maintained an interesting historical myth about a millennium. Western culture married this particular apocalyptic tale with a focus on intervals of 1,000 that any decimal system might be prone to favor. So here we are, engulfed in a millennial madness utterly unrelated to anything performed by the earth and moon in all their natural rotations and revolutions. People really are funny—and fascinating beyond all possible description.

This book, then, focuses on the three great questions that motivate details of millennial madness. My subjects are calendrics, astronomy, and history—not prediction or psychology. I pose, in turn, three of the standard *W* questions. Their resolution should clarify all the major muddles that fuel so much fruitless debate about the millennium in popular media. First, *what* is the millennium after all—and how did the name for a future thousand-year reign of Christ on earth get transferred to the passage of a secular period of a thousand years in current human history? (The connection, both intimate and interesting, forms the subject of Part 1.) Second, *when* does the millennium begin—on January 1, 2000; or on January 1, 2001? (This issue is not nearly so trivial or nitpicking as it might seem, and the nonresolution tells an interesting story about the cultural history of the twentieth century. This section is a revised and extended version of an essay previously published in *Dinosaur in a Haystack*, Harmony Books, 1995. All other material is new and appears here for the first time.) Third, *why* are we so fascinated with calendrical issues about such preferred or "even" transitions as the forthcoming millennial inception (whenever it occurs)? If the universe works like Galileo's grand mechanical clock, regulated by evident mathematical cycles, why does calendrics amount to anything more challenging than simple counting?

We will all end this exploration, I hope, by affirming an amalgam of Einstein's two most famous quotations—both, invoking a metaphorical deity to represent nature's elegant order (or lack thereof). God, indeed, does not play dice with the universe. He is also not at all malicious, though ever so subtle! And, I might add, ever so sly—or do we only see ourselves in a mirror held up to the cosmos?

1

What?

REDEFINING THE MILLENNIUM: FROM SACRED SHOWDOWNS TO CURRENT COUNTDOWNS

OUR NEED FOR MEANING

We inhabit a world of infinite and wondrous variety, a source of potential joy, especially if we can recapture childhood's fresh delight for "splendor in the grass" and "glory in the flower." Robert Louis Stevenson caught the essence of such naive pleasure in a single couplet—this "Happy Thought" from *A Child's Garden of Verses:*

> *The world is so full of a number of things,*
> *I'm sure we should all be as happy as kings.*

But sheer variety can also be overwhelming and frightening, especially when, as responsible adults, we must face the slings and arrows of (sometimes) outrageous fortune. In taking arms against this sea of troubles, no tool can be more powerful, or more distinctly human, than the brain's imposition of meaning upon the world's confusion. This need for meaning becomes especially acute when we fear the accuracy of two great statements fed by Eastern influences into primary documents of Western culture—for these quotations epitomize our suspicion that the cosmos may feature (in our terms) neither sense nor direction, while we humans may inhabit this planet for no special reason and with no goal ordained by nature.

Edward FitzGerald, publishing in the same year (1859) as another revolutionary document filled with challenges to traditional notions of intrinsic meaning, Darwin's *Origin of Species*, freely translated the *Rubaiyat* of the eleventh century Persian poet Omar Khayyam:

> *Into this universe, and why not knowing*
> *Nor whence, like water willy-nilly flowing.*

While the preacher of Ecclesiastes had written, more than a thousand years earlier but with similar doubts about inherently congenial natural order:

> I returned, and saw under the sun, that the race is
> not to the swift, nor the battle to the strong, neither
> yet bread to the wise, nor yet riches to men of

understanding, nor yet favor to men of skill; but time and chance happeneth to them all.

But why invoke such general themes of mental ordering and natural randomness to begin a small book on particular questions about the millennium? I start here because the basic concept of the millennium in Western culture arose from two of the great mental strategies that we use to wrest order and meaning from a recalcitrant world. Moreover, and more particularly, the central shift of meaning that defines our current millennial madness—from millennium as apocalypse to millennium as calendrics—can best be understood as a change of emphasis from one mental strategy to the other.

The First Strategy, Classification

Among the devices that we use to impose order upon a complicated (but by no means unstructured) world, classification—or the division of items into categories based on perceived similarities—must rank as the most general and most pervasive of all. And no strategy of classification cuts deeper—while providing such an even balance of benefits and difficulties—than our propensity for division by two, or dichotomy.

Some basic attributes of surrounding nature do exist as complementary pairings—two large lights in the sky representing day and night; two sexes that must couple their opposing parts to produce a continuity of generations—so we might argue that dichotomization amounts to little more than good observation of the external world. But far more often than not, dichotomization leads to misleading or even dangerous oversimplification. People and beliefs are not either good or evil (with the second category ripe for burning); and organisms are not either plant or animal, vertebrate or invertebrate, human or beast. We seem so driven to division by two, even in clearly inappropriate circumstances, that I must agree with several schools of thought (most notably Claude Lévi-Strauss and the French structuralists) in viewing dichotomization more as an inherent mechanism of the brain's operation than as a valid perception of external reality.

I mention dichotomization as the chief rule of classification because millennial definitions hinge upon our standard (and oversimplified) pairwise divisions for the two most general subjects of all: time and change. For time, Western culture has favored a division between arrows and cycles—or inherently directional versus predictably recurrent sequences of events. (See Mircea Eliade, *The Myth of the Eternal Return*, 1954, for the classic statement, with sources reaching back to Plato and earlier; and my own *Time's Arrow, Time's Cycle*, 1987, for a scientist's perspective on the subject.) For change, we have emphasized the distinction between the gradual and continuous versus the sudden, cataclysmic and revolutionary.

We hold tight to both ends of these dichotomies because each provides part of the psychological comfort needed to survive and prosper in this vale of tears. We need time's arrow to assure us that sequences of events tell meaningful stories and promise hope for improvement. We need time's cycle for an ordered rescue from the fear that history might feature no more than a random and senseless jumble of events without meaning or guidance—just one damn thing after another, in the old cliché. If events recur in predictable ways (as days must follow nights, and new births compensate old deaths), then life includes pattern amidst the flux.

As for time, so also for the dichotomy of change. We need a concept of gradual alteration to sustain hope that what we have built through struggle might persist and even augment—in short, to have some sense of continuity. But we also need the possibility of cataclysm, so that, when situations seem hopeless, and beyond the power of any natural force to amend, we may still anticipate salvation from a messiah, a conquering hero, a deus ex machina, or some other agent with power to fracture the unsupportable and institute the unobtainable.

From these themes of hope and order—and especially from the notion of divine intervention at the end of a determined cycle—we derive one of the most popular and potent beliefs in Western (and many other, perhaps even universal) traditions: *apocalypticism,* defined by *Webster's* as "a doctrine distinguished by the expectation of an imminent end of the present temporal world, the final destruction of the unrighteous in a purging holocaust engulfing the earth, and the resurrection of the righteous to a purified world of bliss." (The word comes from a Greek verb meaning "to uncover" or "to reveal"; *Webster's* particular definition may rely too closely upon a specific Christian myth, but the basic elements of apocalyptic belief surely transcend any particular culture.)

As I shall show in the next section, the original definition of *millennium*—so different from our current madness about calendric transitions at "even" thousand-year intervals—arose from an important feature in the "standard" apocalyptic story of Christian traditions, a truly wild tale from the Bible's last book, chapter 20 of Revelation. The fascinating story of how this concept transmogrified into our current form of millennial madness requires a discussion of the second great mental strategy for ordering our confusing world.

The Second Strategy, Numbers of Ultimate Meaning

The human brain is the most complex computing device ever evolved in the history of our planet. I do not doubt that conventional Darwinian reasons of adaptive advantage underlie its unparalleled size and intricacy. Nonetheless, many of our brain's most distinctive attributes, centerpieces for any concept of a universal human nature, cannot be viewed as direct products of natural selection but must arise as incidental side consequences of the original reasons for such an increase in size. (For example, if I buy a personal computer only to keep the spreadsheet of my family finances, the machine, by virtue of inbuilt structure and quite apart from my intent, can perform a plethora of unanticipated tasks as yet unconceived by any user. The more complex the device, the greater the number of potential side consequences. The human brain is ever so much more powerful than this personal computer.)

The jaws of hell fastened by an angel, from the *Psalter of Henry of Bloise, Bishop of Winchester*. Illuminated manuscript, twelfth century.

Thus, for example, the human brain did not get large so that we could read, or write, or reckon the pattern of solar eclipses—for we developed these skills long after our brain reached its current size. Similarly, I can imagine adaptive value for some aspects of reasoning that might be called arithmetical or calculational. A hunter might want to report the size of a mammoth herd, or a gatherer the dimensions of a field full of tubers. We might require more complex systems to gauge the degrees of bloodline relationship so important to systems of social fealty that might yield Darwinian advantages. But we surely cannot argue that natural selection favored large brains so that we might seek patterns in numerical cycles, and then impart even a "deeper" meaning to these pure and recurring abstractions than we grant to the messiness of natural objects. Yet what foible of our search for knowledge, what intellectual drive of the ages, could be more distinctly human?

In listing the few motivating passions of his life, Bertrand Russell stated that he had "tried to apprehend the Pythagorean power by which number holds sway above the flux." (As the two other chief components of his search for knowledge, Russell then sought "to understand the hearts of men" and "to know why the stars shine"—both also relevant to the questions of millennial madness.)

My argument for the origin of our fascination with numerical regularity closely parallels my claims about our affinity for dichotomous classification. In part, we latch on to numerical regularity, and seek deep meaning therein, because such order does underlie much of nature's patterning. The periodic table, after all, is not an arbitrary human mnemonic, and Newtonian gravity does work by a law of inverse squares. But our search for numerical order, and our overinterpretations, run so far beyond what nature could possibly exemplify, that we can only postulate some inherent mental bias as a driving force. I argued above that this bias almost surely evolved as a side consequence of natural selection, not as a direct adaptation—and must therefore bear a complex and indirect relationship to any concept of utility. Our searches for numerical order lead as often to terminal nuttiness as to profound insight.

The catalogue of numerical schemes seriously proposed as the nature of God or the underlying order of the cosmos would fill a baroque volume of staggering variety. (Nicholas Campion's recent book, *The Great Year: Astrology, Millenarianism and History in the Western Tradition,* Penguin, 1994, presents a good introduction.) Some scholars, divines, and mystics have based their schemes on twos (for our dichotomies), others on threes (for the trinity), others on fours (C. G. Jung's nomination for a primal number), others on fives (for our digits), others on sevens (for the notes of the scale and the planets of Ptolemy's system), others on nines (the square of the trinity) . . . and so it goes.

This theme of number mysticism as a second mental device for ordering nature enters our millennial story through powerful interaction with the first device of twofold classification—particularly with the cyclic side of time's dichotomy and the catastrophic end of change's dichotomy. Imagine the pizzazz gained by any claim for a paroxysmal finale if a sage can penetrate the numerical order of the universe to know exactly how long a current cycle must be—and precisely when it must end!

Millennial thought arises from the linkage of general apocalypticism with a specific numerical theory about the forthcoming end. As stated above, a commitment to cycles based on simple numbers doesn't specify either a duration or an ending time—and nearly anything even mildly plausible has been proposed (and cultishly believed) at one time or another. The particular form of apocalypticism known as *millennialism* or *chiliasm* (from classical words for "a thousand," the first Latin, the second Greek)—the most popular numerical scheme in the history of Christian apocalypticism, though entirely arbitrary with respect to nature, as I argue in the preface of this book—regards the number 1,000 as the hidden basis for both the solution of natural order and the salvation of human souls.

But 1,000 of what, and 1,000 when? The rest of this chapter documents a subtle shift in our primary definition of the millennium—from the duration of a blissful age *following* a forthcoming apocalypse, to the measured passage of a thousand years, perhaps *preceding* the same apocalypse. How and why did we move from the millennium as apocalypse to the millennium as calendrics?

MILLENNIUM AS APOCALYPSE

Millennium does mean, by etymology, a period of one thousand years. This concept, however, did not arise within the field of practical calendrics, or the measurement of time, but in the domain of eschatology, or futuristic visions about a blessed *end* of time. Millennial thinking is embedded in the two major apocalyptic books of the Bible—Daniel in the Old Testament and Revelation in the New. In particular, the traditional Christian millennium is a future epoch that will last for one thousand years and end with a final battle and Last Judgment of all the dead. As described by Saint John in one of his oracular visions (Revelation 20), Satan shall be bound for one thousand years and cast into the bottomless pit; Christ shall return and reign for this millennium with resurrected Christian martyrs. Satan shall then be loosed; he shall team up with Gog, Magog, and a host of other evildoers, for a final battle; Christ and the good guys win, the devils end up in "the lake of fire and brimstone"; *all* the dead are now resurrected and, in a Last Judgment at this true end of time, either rise to live with Jesus or end up in that other, unpleasant place along with most of history's interesting characters.

> And I saw an angel come down from heaven. . . . And he laid hold on . . . Satan, and bound him a thousand years, and cast him into the bottomless pit, and shut him up, and set a seal upon him . . . and I saw the souls of them that were beheaded for the witness of Jesus . . . and they lived and reigned with Christ a thousand years. . . . And when the thousand years are expired, Satan shall be loosed out of his prison, and shall go out to deceive the nations which are in the four quarters of the earth, Gog and Magog, to gather them together to battle . . . and fire came down from God out of heaven, and devoured them. And the devil that deceived them was cast into the lake of fire and brimstone. . . . And

> I saw the dead, small and great, stand before God;
> and the books were opened. . . . And whosoever was
> not found written in the book of life was cast into
> the lake of fire. (Revelation 20:1–15)

The religious and political potency of this vision has resonated throughout our subsequent history. Many statements in the New Testament indicate that Jesus and his initial followers did not expect any long delay in the fulfillment of the apocalypse and the inception of the millennium. Speaking through one of his angels, Jesus states in the last chapter of Revelation (and of the entire Bible): "And he saith unto me, Seal not the sayings of the prophecy of this book: for the time is at hand. . . . Behold I come quickly; and my reward is with me, to give every man according as his work shall be. . . . Blessed are they that do his commandments, that they may have right to the tree of life, and may enter in through the gates into the city" (Revelation 22:10, 12, 14).

The Synoptic Gospels reinforce this theme with a more specific timing. Jesus describes the forthcoming apocalypse in terms similar to John's later account in Revelation (and also to the available Old Testament sources of Daniel and Ezekiel), though without John's flamboyant details: "So shall it be at the end of the world: the angels shall come forth, and sever the wicked from among the just; and shall cast them into the furnace of fire: there shall be wailing and gnashing of teeth" (Matthew 13:49–50). Moreover, Jesus states clearly that the end shall not be long delayed and shall surely occur *within* the lifetime of some people who heard his words:

> Then said Jesus unto his disciples, If any man will come after me, let him deny himself, and take up his cross, and follow me. For whosoever will save his life shall lose it: and whosoever will lose his life for my sake shall find it. For what is a man profited if he shall gain the whole world, and lose his own soul? For the Son of man shall come in the glory of his Father, with his angels, and then he shall reward every man according to his works. Verily I say unto you, There shall be some standing here, which shall not taste of death, till they see the Son of man coming in his kingdom. (Matthew 16:24–28; see also Mark 9:1)

We do not, I think, vitiate the moral value of Jesus' more radical teachings when we properly regard them as rules for action in a corrupt and dying world slated for quick replacement by a blessed age—a new regime

Sinners in Hell, Last Judgement, anonymous. Relief.

that would allocate rewards and punishments according to the character of one's life during the strictly limited tenure of the present order. We might not want to turn the other cheek if bullies and tyrants could look forward to a thousand years of easy domination. And if our worldly gains cannot accumulate for more than a generation or so, while our qualities of soul will determine our future (and eternal) state in a new age so soon to come, then practical reasons of the moment—and not only ethical values for the ages—will favor the calculus of soul over gold.

Jesus' error of timing did not dampen the enthusiasm of apocalyptically inclined supporters, and every subsequent generation has featured millenarian movements. The first Christian version of some significance unfolded only twenty years or so after the Roman suppression of Bar Kochba's revolt finally extinguished Jewish life in Jerusalem, and ended the immediacy of the more secular and messianic form of Jewish apocalypticism. Montanus began to preach in Phrygia (now central Turkey) in about A.D. 156. Aided by two young female disciples, Prisca and Maximilla, Montanus fell into trances and announced the imminent second coming of Christ, as the heavenly city of Jerusalem would descend to earth and establish itself on the plain between the Phrygian villages of Pepuza and Tymion.

Establishing a pattern to be repeated throughout history on many subsequent mountaintops, deserts, valleys, and mesas, the Montanists left their towns (leading to the virtual abandonment of several early Christian communities) and went to the appointed place to await the grand deliverance—which, as usual, and needless to say, never occurred.

However, and also initiating a pattern among true believers that would persist forever after, this spectacular failure of a clear and central prediction did not extinguish the movement, and Montanism remained strong for several hundred years, persisting until the ninth century and even gaining the support of Tertullian, perhaps the most prolific writer among early Christians (Tertullian left the Catholic Church to join the Montanists in 212). Followers admired the asceticism and moral rigor of the movement, and failure of an apocalyptic prediction can always be rationalized with a variety of excuses, from miscalculation to a mixup between metaphorical and literal interpretations.

Meanwhile, as time wore on, and as Christianity became a substantial secular power rather than a persecuted and radical sect, the inevitable backlash occurred, establishing a fundamental contrast that pervades the history of apocalyptic thought. For obvious reasons, established governments, doctrines, and powers must firmly oppose, and actively combat, any prophetic doctrine, and especially any mass movement, centered upon a claim for an imminent and cataclysmic termination of earthly order! Apocalypticism is the province of the wretched, the downtrodden, the dispossessed, the political radical, the theological revolutionary, and the self-proclaimed savior—not the belief of people happily at the helm. What, then, did triumphant Christianity do when its newfound secular success began to press upon the undeniable scriptural authority for apocalyptic expectations?

Two strategies have long prevailed among comfortable establishments that can't deny their own millenarian documents and traditions. First, one can argue that the millennium must, indeed, eventually arrive—but in a future so distant and unknowable that the issue need scarcely influence our daily lives. Second, and more commonly, one can reinterpret the millennium in a metaphorical or allegorical way, and even argue that the blessed event has already occurred. In the classic formulation, virtually canonical in Catholic circles since Saint Augustine's formulation in his early fifth-century masterpiece, *Civitas Dei* (The City of God), the millennium must be viewed allegorically as a spiritual state collectively entered by the Church at Pentecost—the descent of the Holy Ghost to the apostles soon after Christ's resurrection—and fully subject to contemporary personal experience by mystical communion with God. This argument, needless to say, serves a social purpose for a powerful and conservative institution wishing to maintain a status quo of daily influence and to suppress wild theories about actual and imminent ends of the world.

This fundamental sociological division—the key to understanding the power of millennial thinking in Western history—was summarized particularly well, albeit in a highly partisan fashion, by the late seventeenth-century English divine Thomas Burnet (who will figure prominently in the next section of this chapter). As an Anglican priest, a champion of the Reformation, and an anti-Catholic (though not nearly so vehement as many of his famous contemporaries, including Oliver Cromwell and John Milton), Burnet linked millenarian thought to social and religious reform, and then tied the rejection of apocalypticism to support of a comfortable and established order. (I also love the sweep of Burnet's expansive and flavorful seventeenth-century prose style, and I quote from my own copy of his beautiful book.)

I never yet met with a Popish doctor that held [supported] the Millennium. . . . It was always indeed uneasy, and gave offense, to the Church in Rome, because it does not suit to that scheme of Christianity which they have drawn. They suppose Christ reigns already, by his Vicar, the Pope: and treads upon the necks of emperors and kings. And if they could but suppress the *Northern Heresie* [that is, the Reformation], as they call it, they do not know what a Millennium would signify, or how the Church could be in a happier condition than she is. . . . The Church of Rome hath been in prosperity and greatness, and the commanding Church in Christendom, for so long or longer, and hath ruled the nations with a rod of iron. . . . And the Millennium being properly a reward and triumph for those who have come out of persecution, such as have lived always in pomp and prosperity can pretend to no share in it, or benefit by it. This has made the Church of Rome have always an evil eye upon this doctrine, because it seemed to have an evil eye upon her. And as she grew in splendor, and greatness, she eclipsed and obscured it more and more: so that it would have been lost out of the World as an obsolete error, if it had not been revived by some of the Reformation.

Millenarianism drove the most radical of the Reformationists—and only by grasping their firm belief in the imminent end of time can we understand their willingness to engage in militarily hopeless revolt, or to endure the most unspeakable tortures (not that they had much choice) before subsequent execution. Thus, the Anabaptist Thomas Müntzer, convinced that he was living at the very "end of all ages," led the Thuringian peasants in their ill-fated revolt of 1525, and ended up racked and decapitated for his pains. (Martin Luther may have had his radical moments in theology, but he was horrified by the political revolution of the peasants, and he urged that they all be exterminated like dogs, and without mercy—as they were, and by the tens of thousands.)

Millenarian movements have continued on the Protestant fringe (sometimes not so "fringey" in episodes of general enthusiasm or social unrest) and have left their impact upon several major groups (who do not always wish to own their apocalyptic origins). The Hutterite communities of the western United States and Canada, for example, trace their origin to another millenarian German Anabaptist, Jakob Hutter, who was tortured and burned as a heretic in 1536.

The best known, if shortly flaring, millenarian movement in American history reached a climax in New York and Massachusetts during the 1840s, where as many as 100,000 believers followed the apocalyptic message of William Miller, a former army officer and self-proclaimed preacher who announced, based on his reading of Daniel and Revelation, that Christ would return and engulf the world in fiery conflagration sometime between March 21, 1843, and March 21, 1844. When this prophecy failed to materialize (or spiritual-ize), Miller set a later date of October 22, 1844. The uneventful passage of this second Second Coming—known as "The Great Disappointment" in Millerite cir-cles—led to a conference in 1845, devoted to what a later age would call "damage control." Needless to say, many followers had left the fold, for nothing dulls enthusiasm quite so effectively as the spectacular failure of a central prediction.

But nothing can shake the faith of a true believer either. The major group of remaining Millerites argued that Miller had set the right date but had read Daniel incorrectly. God did not wish to end the world on that day, but only to begin his examination of all the names in the Book of Life—a tedious and time-consuming task that would end at some unstated future moment with the appearance of Christ for his millennial reign. Mean-while, the Millerites held that certain practices—partic-ularly the observance of Saturday, the seventh and last day of the week, rather than Sunday—would hasten this process and speed the Second Coming.

The modern Seventh Day Adventists, and other smaller adventist groups, trace their origins to Miller's movement, while not following all his precepts. Charles Taze Russell (1852–1916), founder of the Jehovah's Witnesses—perhaps the largest contemporary and forthrightly millenarian Christian group—was also strongly influenced by Adventist doctrines. The Witnesses regard Satan as currently in control, with secular powers unwittingly under his domination—hence the refusal of believers to salute the flag or undertake military service, the subject of several Supreme Court decisions in our century. They regard Daniel and Revelation as a hidden timetable for human history and await the coming battle of Armageddon and the inception of Christ's reign. Russell himself thought that Christ would begin his "invisible return" in 1874 and stage his true Second Coming in 1914—a good year for assassinations of archdukes and inceptions of world wars, but not for the full blast of Armageddon! However, and once again, the failure of a clear expectation did not derail the passion of true believers, who still ring my doorbell nearly every weekend.

I don't want to make too strict an equation between millenarian belief and social misery or marginality—for human ingenuity, and our self-serving propensities, cut too wide a swath to allow such a potent argument only

Apocalypse of Saint John: Babylon Falls on the Demons (1363–1400),
Nicolas Bataille. Tapestry.

one mode of conceivable action. People in power have also been known to invoke an apocalyptic "gotcha" when the unusual occasion arises. Most notably in recent times, James Watt, Ronald Reagan's unlamented secretary of the interior, a deeply conservative thinker and prominent member of the Pentecostal Assembly of God, stated that we need not worry unduly about environmental deterioration (and should therefore not invest much governmental time, money, or legislation in such questions) because the world will surely end before any deep damage can be done.

Nonetheless, the general correlation of apocalyptic yearnings with earthly poverty and social disenfranchisement surely holds—and extends far beyond purely conventional and western Christian sources. The fusion of Christian millennialism with traditional beliefs of conquered (and despairing) peoples has often led to particularly incendiary, and tragic, results.* In Africa, for

---

*As I read the galley proofs for this book in late March 1997, a tragic event—the suicide of thirty-nine people in the Heaven's Gate cult—made me realize how parochial, even a bit condescending, I had been in writing this statement. I said that some of the saddest results of apocalyptic beliefs resulted from a kind of unholy alliance—when non-Western people selectively chose some Christian bits and pieces and welded them with traditional beliefs to form an unstable and incendiary new doctrine. I should have realized the universality of such propensities and not placed an implied blame upon "others" from cultures so distant from our own. Fully westernized people are

example, several failed revolts and stillborn movements can be traced to an explicitly Christian millenarian inspiration. John Iliffe (*Africans*, Cambridge University Press, 1995) attributes the major defeat of the South African Xhosa people to a natural disaster enhanced by a millenarian response:

> Xhosa tried to incorporate Christian ideas into their cosmology. . . . Mission teaching encouraged this, as did the fact that some Christian ideas had

---

equally capable of fusing apocalyptic Christian bits and pieces with folk myths of our own contemporary culture to form the same kind of destructive and incendiary doctrine. The Heaven's Gate cultists made just such a mixture of traditional millenarianism with American pop culture myths of science fiction in general, and UFOlogy in particular—and the result cost them all their lives. In thinking that their immortal essences resided only temporarily in an earthly body, that the body's individuality must be rigidly suppressed, that their essences would reunite with extraterrestrial higher powers upon their bodies' deaths, and that a space ship awaited behind the tail of the Hale-Bopp comet (which, at this very moment, shines so brilliantly outside my window as I write) to take them "home," they consciously combined Christian millennialism with modern science fiction. Their official statement, prepared before their mass suicide, explicitly said so (and even followed a common pattern in misspelling millennium with only one *n*). They wrote:

> We came from the Level Above Human in distant space and we have now exited the bodies that we were wearing for our earthly task, to return to the world from whence we came—task completed. The distant space we refer to is what your religious literature would call the Kingdom of Heaven or the Kingdom of God. We came for the purpose of offering a doorway to the Kingdom of God at the end of this civilization, the end of this millenium.

radical implications, above all Christian eschatology. Its power was displayed in 1857, at a time of cattle disease and white expansion, when prophets persuaded many Xhosa to kill their cattle and abandon cultivation because their ancestors were to be reborn with finer cattle and drive the Europeans back into the sea. Perhaps one-third of Xhosa died and the Cape Government seized the opportunity to destroy their society, alienating more than half their land and admitting at least 22,150 of them to work in the colony.

Similarly, the most famous of early twentieth-century African revolts, the ill-fated and brutally suppressed 1915 rebellion of John Chilembwe in Nyasaland (now Malawi), had a millenarian basis. Chilembwe had been the servant of Joseph Booth, a fundamentalist missionary who took him to the United States, where he received a degree from a black theological seminary before returning to Africa. Following Chilembwe's execution, his aged mentor Booth lamented this common outcome of an all-too-Christian theme (whatever the inconsistency with other threads of Christian teaching):

Poor kindhearted Chilembwe, who wept with and for the writer's feverstricken and apparently dying child; nursed and fed the father with a woman's tenderness during ten weeks of utter prostration; wept, labored with, and soothed the dying hours of my sweet son John Edward (18 years old) . . . Yes, dear Chilembwe, gladly would I have died by my countryman's shot, to have kept thee from the false path of slaying. (Quoted in Harry Langworth's *The Life of Joseph Booth,* published by CLAIM, the Christian Literature Association in Malawi, Blantyre, Malawi, 1996)

One of the most poignant and tragic of all events in American history, the tale of the last major massacre of Indians by white soldiers—the 1890 Battle of Wounded Knee—arose as a direct, if unnecessary and clearly avoidable, outcome of a fascinating millenarian episode. As R. A. Smith documents in his *Moon of Popping Trees: The Tragedy at Wounded Knee and the End of the Indian Wars* (University of Nebraska Press, 1975), millenarian movements had arisen from time to time among Christianized Indians throughout the United States and Canada. Tavibo, a Northern Paiute from Nevada, had assisted the prophet-dreamer Wodziwob in spreading the Ghost Dance ritual to tribes in California and Oregon during the late 1860s and early 1870s, a movement that had faded after Wodziwob's death in 1872.

Tavibo's son Wovoka (1856–1932) was adopted, at about age fourteen, by the family of a white rancher, David Wilson. Wovoka, now renamed Jack Wilson, became interested in Christianity through the family's nightly Bible readings and general piety. He then studied with Mormon missionaries stationed among the Paiutes and spent some time with the Indian Shaker Church. He developed a potent mixture of Christian apocalyptic beliefs with Ghost Dance lore. Then, during a solar eclipse in early 1889, he experienced a vision of death and had a direct conversation with God, who ordered him to teach the Ghost Dance and its millennial message to his people. Wovoka proclaimed that if the Indians separated themselves from the world, and dutifully performed the Ghost Dance at the appointed intervals, and for the specified time, a millennial renewal would occur: the ghosts of ancestors would return to dwell with the living; the land would be restored to its original cover, richness, and fertility; the white man would disappear; and the buffalo would return.

Wovoka explicitly preached the Ghost Dance as a movement of separation and pacifism; only strict adherence to the appointed ritual—including avoidance of contact, and especially aggression, with whites—could hasten the apocalypse. But given the realities of tension, incomprehension, racism, and recrimination, we can scarcely be surprised that white settlers became distinctly nervous when they observed large groups of Indians abandoning their usual tasks, gathering in central places, and dancing ecstatically for days at a time. The movement quickly spread in all directions, from Texas to the Canadian border, reaching the Sioux in 1890, who added the nerve-racking belief (to whites) that if dancers wore a certain kind of shirt, the white man's bullets could not penetrate.

Many dancers told of their trips to heaven during trances inspired by ecstatic activity. The testimony of Little Wound, chief of the Oglala Sioux, illustrates the fascinating fusion of traditional Christian visions of the millennium with specific Indian themes and grievances, and also with the claim for invulnerability that fanned white fears.

When I fell in the trance a great and grand eagle came and carried me over a great hill, where there was a village such as we used to have before the whites came into this country. The tipis were all of buffalo skin, and we used the bow and arrow, and there was nothing in that beautiful land that the white men had made. Neither would Wakan Tanka let any whites live there. The land was wide and green and stretched in every direction and made my eyes glad.

I was taken into the presence of the great Messiah, and He said these words to me, "My child, I am glad to see you. Do you want to see your children and relations who are dead?" . . . They appeared, riding the finest horses I ever saw, they wore clothes of bright colors that were very fine, and they seemed very happy. . . . The Great Holy made a prayer for our people upon the earth, and then we smoked together using a fine pipe ornamented with beautiful feathers and porcupine quills. Then we left the village and looked into a great valley where there were thousands of buffalo and deer and elk all feeding. . . .

He also told me to go back to my people and say to them that if they would keep on making the dance and pay no attention to the whites that He would shortly come to help them. If the holy men would make for the dancers medicine shirts and pray over them, no harm could come to the wearer; that the bullets of any whites that wanted to stop the Messiah Dance would fall to the ground without hurting anybody, and the person who fired the shots would drop dead.

Two tragedies arising from white misunderstanding of the Ghost Dance movement—the murder of Sitting Bull and the massacre at Wounded Knee—haunt American history. Sitting Bull, who had played a role of disputed importance in the death of Custer at the Little Bighorn in 1876, still led a small group of Sioux, though his importance had been greatly overestimated by local whites. (Sitting Bull had performed in Buffalo Bill's extravaganzas and had become, for racist Americans, a symbol of the recalcitrant, if noble, savage.) Sitting Bull also strongly supported the Ghost Dance movement. The local government agent became alarmed and wrote to the commissioner:

> I feel it my duty to report the present "craze" and nature of the excitement existing among the Sitting Bull faction of Indians over the expected Indian millennium, the annihilation of the white man and supremacy of the Indian, which is looked for in the near future and promised by the Indian medicine men as not later than next spring.

The *Chicago Tribune* then fanned the false flames with a headline for October 28, 1890, based on this agent's letter:

TO WIPE OUT THE WHITES
What the Indians Expect of the Coming Messiah
Fears of an Outbreak
Old Sitting Bull Stirring Up the Excited Redskins

The commissioner decided to arrest Sitting Bull and judged that the Indian Police could do the job most efficiently and diplomatically. But high tension, combined with the usual misinterpretations and avoidable provocations, turned a peaceful mission into a carnage, as gunfire broke out on both sides, leaving six policemen and eight of Sitting Bull's supporters dead, including the old chief himself.

White nervousness about the Ghost Dance also led the government to a tragic decision to round up Chief Big Foot's independent band of Lakota Sioux and bring them into confinement on the reservation. The mission proceeded with great tension. Neither Big Foot nor the army commander sought any trouble, and both tried to calm the rising tempers, fanned primarily by hotheaded young men on both sides—the army recruits filled with stereotypes and fears, the Indians burning with anger and legitimate grievances. If Big Foot had not been too ill to lead, and if the soldiers had been seasoned veterans rather than scared neophytes, the unjust mission, distasteful to both sides, would no doubt have been quietly accomplished as planned. But the usual set of banal, utterly unheroic, small, and avoidable events occurred, and—just as with Sitting Bull but at greater scale—panic and gunfire broke out at Wounded Knee Creek, South Dakota, on the morning of December 29, 1890. Government soldiers fired directly into fleeing groups of Indians. When the panic eased and the smoke cleared, 30 white soldiers had been killed, while 84 Indian men and 62 women and children, also lay dead. The ghost shirts had not worked.

In summary, Richard Landes, a history professor at Boston University, director of their Center for Millennial Studies, and a specialist on the millennial movements of medieval Europe, offers a powerful argument for the great importance of apocalypticism in history. Two major reasons, common to almost all millenarian episodes, govern the general significance. First, the certainty felt by true believers about an imminent termination to the current order leads them to break social traditions, roles, and boundaries that they would never dream of fracturing in normal times. (Why worry about slave and master, if all good people will soon rise to common glory with Christ? Why kowtow to local lords or obey unjust regional ordinances, if the Son of Man comes quickly, and all will soon serve God alone?)

Second—and citing the preeminent empirical regularity of this recurring history—millennial expectations *always* fail, and their movements are left with a host of radical social practices that must now be reconciled with an unexpected and continuing life on the present earth. These novelties, now transferred from designs for the blessed millennium to devices for potential reform of the current order, often become a motor for major disruptions and transformations on the complex pathways of human history. Jesus said that "one jot or one tittle shall in no wise pass from the law"—"till all be fulfilled," and "till heaven and earth pass." But heaven and earth stayed put, while the energy of unfulfilled millennial expectations dislodged the jots and tittles of every "permanent" holy text, and altered the jobs and titles of every "eternal" social status—with consequences that have often led either to genocide, or to liberation.

MILLENNIUM AS CALENDRICS

The Stoicks tell us, When the Sun and the Stars have drunk up the Sea, the Earth shall be burnt. A

Detail from *The Last Judgement*, Fra Angelico (1387–1455).

very fair prophecy: but how long will they be a
drinking?

> The Reverend Thomas Burnet
> *Telluris theoria sacra* (The Sacred
> Theory of the Earth), 1691

Why Make the Switch at All?

Millennial disappointment—from the failure of Jesus'
initial prediction for an apocalypse within his own generation, to the latest slinking away of the faithful from
the most recently designated mountaintop—must provoke one anguished question above all others: "If not
now, then when?" (to cite a famous Jewish proverb for a
different purpose).

The last section documented the original meaning of
the millennium in Christian and Western history—a
specific fable about Christ's *future* reign for a thousand
years on earth, following an apocalyptic destruction of
the present order. But as the year 2000 approaches, the
primary definition of *millennium* has shifted to a quite
different and primarily calendrical meaning—the completion of a secular period of a thousand years in
human history, particularly when packaged between

beginning and ending years with nice, clean, "even" designations, like 1000 and 2000.

Are these two notions—the millennium as apocalypse and the millennium as calendrics—related at all? Or are we only using, albeit with etymological justification, the same word for two different concepts sharing a merely coincidental concern with durations of a thousand years? (After all, if *simile* tells me to "make way" for a Swahili dignitary, but only recognizes an English figure of speech, then why can't *millennium* have two disparate meanings with only an incidental link to the number 1,000?) Fortunately for the goals of this book (which would immediately lose all reason for existence otherwise), and for the sake of a good tale in general, the two usages enjoy a sensible and intimate historical linkage—all proceeding from the cardinal question that opens this section: "If not now, then when?"

The basic reason for switching from a description of the future to a counting in the present stems from the failure of this expected future to materialize. If you invite ten people to dinner on Saturday and nobody shows up, then you should check your calendar for the most probable explanation. They may all have died on the road, or come down with the flu and forgotten to call. However, I'd be willing to make a substantial bet that you misremembered the appointed date—and that your guests will all show up at the designated hour, but on the following Saturday. Similarly, if you just *know* that the millennium must come, decide on next Thursday as the due date, and end up that night as a bridesmaid on the stroke of midnight after a long wait, what will you assume? Either you were in error about the millennium happening at all—a possibility simply too brutal to contemplate for many people—or you got the date wrong (an unhappy circumstance to be sure, but ever so much better than the alternative).

Your initial concern may have been preparatory: What shall I do on that great gettin' up morning? Through which of the twelve gates of the city shall I enter? But your new question has to be calendrical: All right, so I was wrong about Thursday. But when will the millennium come?

This obvious rationale for a switch to calendrical issues provides only a small part of the answer to our primary question: Why change the definition of *millennium* from an apocalyptic description to a calendrical interval? That is, we can easily grasp the new concern for calendrics, but why in heaven's name should we have any preference, or any concern at all, for the number 1,000? The millennium will bring us one thousand years of future bliss whenever Christ decides to make his delayed entrance, but why should our revised estimate for the Second Coming invoke any interval of one thousand anythings? The duration of future pleasure bears no intrinsic or necessary relationship to the agony of current waiting.

History clearly affirms this logical verdict. Hardly any question has yielded a greater variety of answers, following calculation by disparate rationales under different assumptions, than the forthcoming inception of the millennium (which did not arrive, as initially promised, during Jesus' own generation). Millennial expectations throughout the ages have been generated from all manner of systems, some numerical, some hermeneutic, some visionary, some supposedly empirical and scientific, and some downright hallucinatory. Cyclical theories of repetitive time have been favored, with the millennium following the completion of a cycle. Thomas Burnet defended the partisans of cyclicity in his millennial treatise of the early 1690s, *The Sacred Theory of the Earth:* "Their revolution to the same state again, in a great circle of Time, seems to be according to the methods of Providence; which loves to recover what was lost or decayed, after certain periods: and what was originally good and happy, to make it so again."

Almost any possible numerical basis has been advocated for determining the length of a worldly cycle and the subsequent initiation of the millennium. Many adepts favored a division by two, as the birth of Jesus initiated a second age that would repeat (symbolically) all the events of the Old Testament until the completion of the rerun marked the end of earthly time, and the dawn of the Second Coming. The theory of the Third Disposition, promoted by the most famous millenarian thinker of medieval times, the twelfth-century Italian mystic and biblical philosopher Joachim of Fiore, provided a prototype for popular theories of threefold cycles based on the trinity. (Joachim divided earthly history into a cycle of three "dispositions" representing ages of the father, son, and holy spirit.) Many other thinkers preferred a fourfold cycle based on the Four Empires of the apocalyptic chapters of the Book of Daniel. Still others advocated a fivefold division based on the five sequential political societies of Plato's Great Year. The most potent and secular of quasi-apocalyptic movements in our own times, the earthly millennium of communism promised by Marx's theory of ineluctable stages in the history of social organization, promoted a division by six—primitive communism, slavery, feudalism, capitalism, socialism, and communism (with the last stage read as an improved return to an original blessing). Saint Augustine preferred a Great Week of seven historical phases, all of different but predictable length.

Obviously, with such diversity in the bases of judgment, intervals of a thousand years could enjoy no inherently favored status. Thus, the millennium has been predicted and expected at almost any time, depending on the system in favor. Obviously, with Thomas Müntzer advocating 1525; William Miller, 1844; Wovoka, 1890; and John Chilembwe, 1915; the year 1000 or 2000, and intervals of 1000 in general, could claim no special preference.

Why Favor Intervals of One Thousand Years?

Within this maelstrom of diversity, however, a particular argument arose during the earliest days of Christianity and gained strength forever after to grant the number 1,000 a highly preferred status as a favored figure in the history of calendrical calculations for the millennium. This argument did link the thousand-year future duration of the millennium with the passage of thousand-year intervals of earthly history—so the two apparently disparate usages do become fused after all. The millennium as apocalypse does lead to the millennium as

The Last Judgment, anonymous, Bologna, fourteenth century.

calendrics—but only through an argument steeped in symbolism.

After this long buildup, the classic argument for link-ing the apocalyptic and calendrical millennium may seem awfully weak and disappointing—for the junction requires a symbolic interpretation that will probably strike most of us today as fatuous and far-fetched. Has so much ever been based on so little? But our secular todays provide no basis for judging the apparent strength and good sense of an argument to our more spiritually inclined forebears—for whom a symbolic link often seemed both brilliantly illuminating and entirely conclusive. Or so they said, at least—and I think we must take them at their word. It may be ours to reason why (as we try to understand); but it is *not* ours to deny the satisfaction felt by our forebears because we no longer credit a style of argument once equated with our modern regard for empirical science as a pathway to truthful answers about the natural world.

The classic argument is "only" an analogy, and we now tend to regard analogies as, at best, "cute" and, at worst, "misleading." We certainly judge analogy as the poorest relative (if not an entirely foreign interloper) in a family of useful approaches ruled by the twin mon-archs of irrefutable internal logic and ascertainable sen-sory data. But if we lived in a world where God made every item, from molecule to Milky Way, for a purpose accessible to human ingenuity, then we might develop a different theory of proof and meaning. If all natural objects were created as intended parts of an integrated and completed whole—and if this entirety enfolds a meaning that may be difficult to ascertain (for God works in mysterious ways), but surely holds the secret to joy and understanding if we can only find the key—then a search for "deep significance" in interrelationships among superficially disparate parts may become our method of choice. Analogy may then stand forth as our most valuable tool—for how else can we link the sand grains in the desert to the stars in the sky—and not just as a funny frill for banter at next Saturday's party (when those ten guests will finally visit).

As divines and scholars searched the New Testament for clues to a revised date for the millennium, they focused upon chapter 3 in the Second Epistle of Peter—a letter to the faithful about current millennial disappointments and future expectations. Peter begins by acknowledging the doubt generated by the nonoccurrence of the apocalypse at the expected time: "There shall come in the last days scoffers, walking after their own lusts, and saying, Where is the promise of his coming? for since the fathers fell asleep, all things continue as they were from the beginning of the creation."

But Peter then reminds us that all has not been calm and uneventful since the beginning, for God had destroyed the early earth by water in Noah's flood: "For this they willingly are ignorant of, that by the word of God the heavens were of old, and the earth standing out of the water and in the water: Whereby the world that then was, being overflowed with water, perished." Moreover, several biblical prophecies suggest that the next destruction shall be by fire: "But the heavens and the earth which are now, by the same word are kept in store, reserved unto fire against the day of judgment and perdition of ungodly men."

But when shall this day of judgment arrive, and when shall the promised millennium begin? Peter does not give a direct answer, but rather, in the next verse of chapter 3, makes the symbolic argument by analogy that would set the course of millennial debate forever after: "But, beloved, be not ignorant of this one thing, that one day is with the Lord as a thousand years, and a thousand years as one day" (2 Peter 3:8).

Peter therefore gives an oracular answer rather than a particular date, but at least he cites the familiar symbol of a friendly oracle, not an idiosyncratic and novel voice—for the equation of our thousand with God's unity sets a common theme in the Old Testament, particularly in the celebrated words of Psalm 90: "For a thousand years in thy sight are but as yesterday when it is past, and as a watch in the night." (The same psalm also contains the classic line for a linkage between counting and understanding: "Teach us to number our days, that we may apply our hearts unto wisdom.")

With this guiding equation, we can reason our way to a duration for earthly time that must herald the millennium. The Book of Revelation says that the first age of postapocalyptic bliss (the millennium) will last for a thousand years. We know that any period of a thousand years represents only a day for God. We also know that God created the world in six days, and then rested on the subsequent seventh. Therefore, by symbolic comparison, the world's history will unfold for six thousand years to a point of completion for ordinary earthly time (comparable with God's fulfillment of the initial creation), and will then enter the seventh and final thousand-year period of millennial bliss (comparable with God's day of well-deserved rest following his Herculean labors). The history of the earth, therefore, must span exactly seven thousand years—symbolizing God's seven days of creation (six of work and one of rest, corresponding with six thousand years of human pain followed by one thousand years of millennial harmony)—before *Tuba mirabilis* (the wondrous trumpet) of the Last Judgment announces a true and ultimate finale.

The future thousand-year duration of the millennium does therefore specify a calendrical count by thousands for the appointed length of human history, and for knowing the crucial moment of termination by the Second Coming of Christ. This standard argument—surely the most familiar, and most widely accepted, calendrical theory for the millennium throughout Christian history—dates at least to the early fourth-century writings of the church father Lactantius, who stated in his principal work, the *Divinae institutiones* (Divine Precepts):

> Plato and many others of the philosophers, since they were ignorant of the origin of all things, and of that primal period at which the world was made, said that many thousands of ages had passed since this beautiful arrangement of the world was completed; . . . But we, whom the Holy Scriptures instruct to the knowledge of the truth, know the beginning and the end of the world. . . . Therefore let the philosophers, who enumerate thousands of ages from the beginning of the world, know that the six thousandth year is not yet completed, and that

when this number is completed the consummation must take place, and the condition of human affairs be remodelled for the better, the proof of which must first be related, that the matter itself may be plain. God completed the world and this admirable work of nature in the space of six days, as is contained in the secrets of Holy Scripture, and consecrated the seventh day, on which He had rested from His works. . . .

Therefore, since all the works of God were completed in six days, the world must continue in its present state through six ages, that is, six thousand years. For the great day of God is limited by a circle of a thousand years, as the prophet shows, who says, "In Thy sight, O Lord, a thousand years are as one day."

At the end of the seventeenth century, almost 1,300 years later, the Reverend Thomas Burnet presented the same argument in his millenarian treatise on both human and geological history, *The Sacred Theory of the Earth:*

It is necessary to show how the Fathers grounded this comparison of six thousand years upon scripture. 'Twas chiefly upon the Hexameron, or the Creation finished in six days, and the Sabbath ensuing. The Sabbath, they said, was a type [symbol] of the Sabbatism [the Millennium], that was to follow at the end of the world; and then by analogy and consequence, the six days preceding the Sabbath must note the space and duration of the world. If therefore they could discover how much a day is reckoned for, in this mystical computation, the sum of the six days would be easily found out. And they think, that according to the Psalmist and St. Peter, a day may be estimated a thousand years; and consequently six days must be counted six thousand years for the duration of the world. This is their interpretation, and their inference.

Burnet then acknowledged the "essential weakness" in principle for all such arguments by allegory and analogy. (He was, after all, a contemporary and pal of Newton, and the modern age, with new criteria for the validity of arguments, was dawning.) Yet Burnet could find no factual problem with this traditional view, and he therefore offered his warm support: "We may be bold to say, that nothing yet appears, either in nature, or scripture, or human affairs, repugnant to this supposition of six thousand years, which hath antiquity and the authority of the Fathers on its side."

As a final point, this allegorical comparison of divine days and human ages also secured a special status for calendrical millennia as preferred units of division and counting. If human history had a fixed duration of six thousand years, and if each millennium of this totality symbolically replayed a discrete day of God's creative work at the inception of time, then millennia became "atoms" of historical time—the basic and indivisible units of our reckoning. Any good and comprehensive theory designates a basic unit by the logic of its explanatory structure—and such units are therefore "theory-bound," and not entirely (sometimes not even substantially) matters of objective observation. Atomic theory gives us the periodic table for units, or elements, of matter. Particle physics gives us quarks, charms, flavors, or whatever comes next in a changing field, for building blocks at the smallest scale. Evolution gives us species for the natural parsing of organisms. And the allegory of God's days gives us millennia for the fundamental divisions of sequential time.

At the same time, since theories represent such interesting and complex mixtures of empirical reality and human preference, and since theories are often so historically contingent and so remarkably wrong, I must

Detail from *The Last Judgement* (1443), Rogier van der Weyden. Altarpiece, center panel.

also remind readers (as the introduction to this book stresses) that nature's genuine astronomical cycles (days, lunations, and years) recognize no division by thousands at all, however much our religious history, and decimal mathematics, may legitimately choose to favor such a criterion of counting.

Why Grant Significance to "Even" Years with Three Zeros?

The allegorical comparison of God's days with human millennia only provides half an answer to the burning practical question that inspired this entire exercise: Just when, exactly, will the apocalypse unfold and the millennium begin? We can ascertain that this big bang in earthly time will occur after the completion of six thousand years, marked as six ages of one thousand years each. But we cannot tell when the six thousand years will finish until we also know when time began! Give me a beginning point *and* a duration—and I can specify the end with precision.

If the idea of a six thousand year duration won majority support (and could at least be easily understood by those who favored different systems for calculating the length of historical time), the second essential ingredient—the fixation of a starting date—provoked no end of dispute and never led to any consensus. Various prophets of the millennium could therefore continue to hawk their different moments.

Sextus Julius Africanus (ca. 180–250), a Roman official and early Christian scholar, developed the first popular system for time's ending based on the twin precepts of a specified beginning and a six thousand year duration. In the first universal chronology written from a Christian perspective, the extensive *Chronographiai* of 221, Sextus argued that five thousand years had passed from the creation to the Babylonian captivity of the Jews, and an additional five hundred from then to the birth of Christ. This countdown left only five hundred years until the appointed end of time. Sextus therefore announced that the millennium would begin in A.D. 500—a date sufficiently distant to preclude any embarrassing disproof during his own earthly existence, but soon to fail in the ultimate court of appeal, as the year 500 passed without any cataclysm worthy of note.

The Christian world then got all het up over an alternative calculation that foresaw the millennium in 800 or 801. The specified year did become a milestone in European history, but the coronation of even so regal a figure as Charlemagne must rank as small potatoes to the anticipated, but unfulfilled, inauguration of the Kingdom of Christ.

Much later now, as the year 2000 approaches, thoughts turn once again to the apocalypse—though mainly with wry amusement or scholarly interest in our unabashedly secular age, rather than with quaking fear or fervent anticipation. The popular impression that apocalyptic yearnings should peak in years with three zeros will—if substantiated—forge a strong and final bond between the millennium as apocalypse and the millennium as calendrics. We must therefore ask whether the old belief in a six thousand year duration of ordinary earthly time, followed by an additional thousand years of millennial bliss before the Last Judgment, also included any preference for years with three zeros as points of transition between millennial ages— particularly for the cardinal moment of the Second Coming and cataclysmic passage from secular to divine government.

As an empirically minded scientist, let me back into this key issue with *the* single source of testable information that limited time has made available to us. The theory of a six thousand year duration arose early in the history of Christianity, and only one millennial transition has occurred since then: the year 1000. As many people already know—and as many more will soon learn from the growing literature inspired by our forthcoming millennial moment—the issue of whether a so-called "panic terror" swept Christian Europe in the year 1000 has provoked a major debate among professional historians for quite some time.

This subject has spawned an enormous literature, both technical and popular, and spanning the full range of opinion from virtually complete denial (Hillel Schwartz, *Century's End,* Doubleday, 1990) to absurdly uncritical acceptance (Richard Erdoes, *AD 1000,* Harper and Row, 1988), all balanced by the nuanced intermediacy of a consummate professional (Henri Foçillon, *The Year 1000,* Frederick Ungar, 1969).

Foçillon allows that apocalyptic stirring certainly occurred, at least locally, in France, Lorraine, and Thuringia, toward the middle of the tenth century. But he finds strikingly little evidence for any general fear surrounding the year 1000 itself—nothing in any papal bull, nothing from any pope, ruler, or king.

On the plus side, one prolific monk named Raoul Glaber certainly spoke of millennial terrors, stating that "Satan will soon be unleashed because the thousand years have been completed." He also claimed, though no documentary or archaeological support has been forthcoming, that a wave of new church-building began just a few years after 1000, when folks finally realized that Armageddon had been postponed: "About three years after the year 1000," wrote Glaber, "the world put on the pure white robe of churches."

Glaber's tale provides a striking lesson in the dangers of an idée fixe. He was still alive in 1033, still trumpeting the forthcoming millennium—though he admitted that he must have been wrong about Christ's nativity for the beginning of a countdown, and now proclaimed that the apocalypse would surely arrive instead at the millennium of Christ's Passion, in 1033. He read a famine of that year as a sure sign: "Men believed that the orderly procession of the seasons and the laws of nature, which until then had ruled the world, had relapsed into the eternal chaos; and they feared that mankind would end."

My own position had favored skepticism until I attended an international conference devoted to this subject ("The Apocalyptic Year 1000," Boston University, November 3–5, 1996). There I learned how rich and complex this debate has become. First of all, the "panic terror" has long been a political football in historical circles. French romantic historians of the early nineteenth century loved the legend and constructed an elaborate set of arguments to justify a potent and widespread apocalyptic episode. But positivist historians of the subsequent Third Republic, imbued with the rationalist spirit of the late nineteenth century, adopted an opposite and skeptical attitude that has dominated the profession to the present day.

Medieval historian Richard Landes, convener of the conference, convinced me that sufficient evidence now exists to support at least a modest claim for substantial millennial stirring, especially in peasant and populist strata of society—the very groups that leave so little historical record of their potent concerns, all the more so in this distant age before printing. At least I am convinced that my strongest reason for skepticism can be laid to rest. I had not even been persuaded that a year 1000 existed in the consciousness of most people at the time. Our current B.C.–A.D. system for counting years did not arise until the sixth century (see Part 2), and I thought that this scheme had made little headway into popular consciousness by the year 1000. But Landes and others have shown that the famous chronologies of the Venerable Bede, that redoubtable eighth century English cleric and scholar, had been copied extensively and widely distributed to almost canonical use among ecclesiastical timekeepers throughout Europe. Bede followed and popularized the B.C.–A.D. system. Through his works, the advent of the year 1000—and its millennial implications—had probably diffused to all social classes.

This tale of the year 1000 establishes the last link in our progression from the original millennium as a future epoch of a thousand years to our current use of the same word for ends of thousand-year periods centered on nice round years with three zeros. One common character anchors this shift in meaning: Jesus Christ himself. The original millennium specified the length of his reign *after* the Second Coming. To forecast this blessed event, early Christians postulated a six thousand year duration for ordinary earthly time, parsed as six periods of a thousand years apiece. To make these calendrical millennia turn at years with three zeros, fraught (like the forthcoming year 2000) with such earnest and worldwide anticipation, we must center our system for counting years on an event that supposedly occurred at one of these "nice round" moments.

Our current system of counting uses the traditional birth of Jesus as such a centering point. The architects of our calendar counted backward from this beginning, in packages of millennia B.C., until they reached the creation of the world. They then counted forward, in packages of millennia A.D., to fulfill the six thousand years of human history, and to specify the apocalypse of the Second Coming. It all makes sense. A mathematically inclined God, mindful of the allure of cycles and numerical repetitions for the lovable and fallible creatures that he had crafted in his own image, would surely have incarnated his only begotten son at a crucial turning in the cosmic cycle of thousands.

Only one question now remained—the most practical and portentous of all: At which thousand-year turning had Jesus been born? How many of the six possible millennia had preceded his birth, and how many would be left for our future? The people who succumbed to the panic terror of the year 1000 thought that five millennia had preceded the birth of Jesus, and that the apocalypse must therefore arrive at the next turning.

Once again, and as always in the history of apocalyptic thought, the appointed time passed and the earth endured. Traditionalists therefore revised their theory in the obvious minimal manner: four thousand years must have elapsed between the creation and the birth of Jesus—and the current world could therefore endure until the year 2000.

The beginnings of modern historical scholarship in the seventeenth century provide a final chapter to our story. Creation in 4000 B.C., and destruction in A.D. 2000, could be validated by symbolism and allegory. But why not seek corroboration from the data of human history? The Bible and other historical documents presented the chronology of human life. Why not count backward from the birth of Jesus, through the duration of Roman and Near Eastern empires, the reigns of the kings of Judah and Israel, the ages of the patriarchs (including Methuselah's maximal 969 years), and the week of creation, to see if a beginning in 4000 B.C. could be squared with the historical record?

Available documents had already made estimates "in the right ballpark," thus auguring well for the success of a more rigorous application. Using somewhat different systems of reckoning, the Hebrew Bible had set creation at 3761 B.C., while the Septuagint (the Greek Bible, translated by the Jews of Alexandria) favored 5500 B.C. Several pre–seventeenth century scholars had also tried their hand, with similar results. The Venerable (and apparently ubiquitous) Bede had calculated 3952 B.C., a figure tantalizingly close to the preferred date of 4000.

But the seventeenth century marked the golden age in this enterprise of scouring historical records to set the limits of time. We tend to scoff at these efforts today, branding them as the last holdout of an unthinking and anti-intellectual biblical idolatry. I will not, needless to say, defend the enterprise for any factual acuity. These scholars made a crucial error in choosing to regard the

Bible as literally true. Since the "week" of creation is too short by several orders of magnitude, the calculated dates obviously bear no relationship to the true extent of geological history! But we cannot fairly invoke our present knowledge to castigate past scholarship based on different and honorable (if incorrect) premises. The calendrical counters of the seventeenth century included the brightest and most learned scholars of the time. Their efforts marked a high point in traditions of humanism, for these scholars committed themselves to an exclusive use of data and reason (though we now view their data as insufficiently accurate, and their reasoning as crucially misguided on the fundamental issue of biblical literalism).

Archbishop James Ussher, the Anglican Primate of All Ireland (an ecclesiastical title for a leader among bishops, not a zoological designation for a monkey's uncle), published the most famous of all chronologies in 1650: *Annales veteris testamenti a prima mundi origine deducti* (The Annals of the Old Testament, Deduced from the First Origin of the World). Ussher set the moment of creation at a day that would live in both infamy and memory—4004 B.C. (at noon on October 23). Let no one saddle the good archbishop with any charge of imprecision!

Ussher's figure lies so tantalizingly close to the

*The Opening of the Fifth and Sixth Seals, the Distribution of White
Garments Among the Martyrs and the Fall of Stars* (1498), Albrecht
Dürer. Woodcut from The Revelation of Saint John.

expected date of 4000 B.C. Only one tiny question of reconciliation remains—and we may bring this inquiry to a close: Where did Archbishop James Ussher find those four little extra years, and why did he feel compelled to include them? Did the biblical dates just add up to this sum? Did the good archbishop then decide, after so many years of such concentrated labor, that calendrics could work like horseshoes—one of the few human enterprises, or so the saying goes, where "close enough" counts? Or do the four extra years arise for an interesting and principled reason that can round out our story?

Happily, the more interesting alternative applies. As I shall show in Part 2, the sixth-century inventor of the B.C.–A.D. system made an unfortunate little error in setting the birth date of Jesus at the crux of his transition. Herod, you see, died in 4 B.C. So if Herod still ruled at the birth of Jesus—and think of how many good stories must disappear if he did not (the slaughter of the innocents, the return of the three magi to their own country)—then Jesus must have been born in 4 B.C., if not earlier. Ussher therefore tacked these four additional years onto his chronology—for theory dictated that exactly four thousand years must pass from the creation to the birth of Jesus, thus setting the beginning of the world at 4004 B.C.

Ussher fully accepted the standard view that exactly six thousand years must pass between the creation and the Second Coming. He performed his calculations partly to determine when the world must end, and in the hope that this blessed millennium would arrive soon enough to fuel human hope, but at a sufficient distance to spare his own life and power. Ussher was also a partisan of the switch from millennium as apocalypse to millennium as calendrics. That is, he advocated the notion that earthly time should be counted in units of 1,000, and that each millennial transition should be marked by a great historical event to signify the overall beauty and internal logic of God's system—with the last moment, the inception of the true millennium, at exactly six thousand years from creation.

Ussher argued that Solomon had completed his temple at the halfway point of 3,000 years, and that Jesus must appear exactly a thousand years later at 4000 A.M. (*Annus Mundi,* or "year of the world"). Moreover, Ussher followed the old medieval theory of types that viewed each story in the New Testament as a symbolic replay of an Old Testament event—so that time's six thousand years formed two great and coincident cycles, with the completion of the second cycle marking the end of business as usual and the advent of the millennium. Thus, although we may now view the bases of comparison as far-fetched or even risible, Mary, when pregnant with Jesus, served as the type of Moses' Burning Bush—because both held the fire of God within themselves yet were not consumed. And the Resurrection of Jesus must replay the deliverance of Jonah from the whale—because both men were buried in death and darkness, but exited from their tombs on the third day. For Ussher, the birth of Jesus represents the type for the completion of the Temple—the establishment of the new and old orders. A neatly numerical God, working within his six thousand year framework, would surely place these events at two successive millennial cruxes, separated by a thousand years.

So four thousand years must separate the creation from the birth of Jesus, who appeared on earth at exactly 4000 *Annus Mundi*. But that nasty little problem about Herod's death had thrown God's elegant reckoning four years out of kilter with the erroneous, but official, B.C.–A.D. system that regulated the secular calendar. Thus, on our flawed calendar, Jesus was born in 4 B.C., and the world—necessarily created exactly four thousand years before—began in 4004 B.C. Ussher wrote (and I quote from my own copy of this amazing book):

The true nativity of the Savior was full four years before the beginning of the vulgar Christian era, as is demonstrable by the time of Herod's death. For according to our account, the building of Solomon's Temple was finished in the 3000 year of the World, and in the 4000 year of the World, the days being fulfilled in which the Blessed Virgin, Mother of God, was to bring forth Christ himself (of whom the Temple was a type) was manifest in the flesh, and made his first appearance unto man: from which four years being added to the Christian era, and as many taken away from the years before it, instead of the Common and Vulgar, we shall obtain a true and natural Epocha of the Nativity of Christ.

Ussher's large folio volume represents an immense labor of calculation and scholarship (requiring knowledge of Latin, Greek, and Hebrew). You can't simply spend a rainy afternoon counting the begats in the Bible, for gaps and ambiguities abound, and the record is incomplete in any case—for the chronology of the Old Testament ends with the books of Ezra and Nehemiah in the fifth century B.C., and the New Testament doesn't pick up again until the time of Jesus. Thus, one has to move laterally from the biblical record into the historical documents of other societies (particularly to Babylon, where biblical and Babylonian records can be correlated for the captivity of the Jews), then forward to Roman history, and back again to the New Testament.

We could be uncharitable and suspect that all Ussher's work amounts to little more than an elaborate scholarly smokescreen for a preconceived conclusion. Ussher "knew" that the earth must last for precisely six thousand years, and that Jesus must have been born in exactly 4000 *Annus Mundi*—so didn't he just jiggle and poke the data until the dates came out "right"? Perhaps, but I don't think so—or at least I am confident that Ussher proceeded with honorable intentions and methods (even if his preconceptions unconsciously affected his procedures). All scholars must begin with a theory in mind, and work to test—and to reject if necessary—an original preference. Ussher knew what he wanted, but he began with no guarantee that the data would validate his desires.

To be fair to the cynics—who stress the implausibility of getting real data to match an admittedly nonsensical, and floridly false, theory with such precision—Ussher must have massaged all the gaps and ambiguities to his advantage. The data contain enough "slop," enough missing intervals where a scholar must extrapolate across a gap, to provide a great deal of "play" and plasticity for squeezing information into expectations. But the same data also impose strong constraints upon a vivid imagination. The actual information must come pretty darned close to four thousand years for the distance between creation in a literal Bible (where the days of God's first active week can have no more than twenty-four hours) and the birth of Jesus. If the stated lifetimes of the patriarchs, and the given reigning times of the kings, added up to ten thousand or two thousand years, then this enterprise would be cooked, and a system of allegorical reasoning would have to be invented by those who still "knew" when the world must end. So let's be kind to Ussher and honor his substantial labor. I don't doubt that he read all questionable points in his favor, but he did count, and labor, and read, and ponder, year after patient year.

So the world must end, and the millennium begin, at exactly 6000 *Annus Mundi*, precisely two thousand years after the birth of Jesus. Well, the year 2000 lies just around the corner, so maybe we should be preparing by learning how to gnash our teeth, and by inventing some really good (and noncarcinogenic) asbestos substitute for the forthcoming fire and brimstone. But wait a minute. Jesus was born in 4 B.C.—so 6000 *Annus Mundi* has already come and gone, precisely on October 23, 1996, by Ussher's chronology. What happened?

Well, something suggestive did transpire on that date. George Burns once said, with undeniable justice, that the victory of the New York Mets in the 1969 World Series constituted the first indubitable miracle since the parting of the Red Sea. So if God, by his common touch, now signals us through crucial events in our secular culture, October 23, 1996, did feature a prominent miracle. The New York Yankees, dangerously behind, two games to one, in their World Series with the powerful Atlanta Braves, were hopelessly in arrears, six to three, with only five outs left in the eighth inning of the crucial fourth game—where a loss, and a consequent three to one deficit, would have sealed their fate. The Yankees won that game in one of the most miraculous and improbable comebacks in the history of sport. So, on the eminently reasonable assumption that God is a Yankee fan (and both a kindly and inscrutable figure as well), He may have used 6000 *Annus Mundi* to send a signal and solicit our earnest preparation before He runs out of reasons for delay and must ring down the truly final curtain on earthly business as usual.

But wait one really last minute. As the next section will show, October 23, 1996, was not 6000 *Annus Mundi* after all! Dionysius Exiguus, that pesky sixth-century monk who also committed the four-year blunder about Jesus' birth, made another portentous mistake in establishing the B.C.–A.D. system. He didn't include a year zero in the transition—the reason, as we shall see, for the perennial debate about whether centuries begin with the '00 or the '01 year, and whether the new millennium arrives in 2000 or in 2001. Thus, thanks to this missing year, 6000 *Annus Mundi* will occur on October 23, 1997, by Ussher's chronology!

Whew! for I am writing this essay in January 1997—so I still have a little time to prepare (and I better watch out, and better not pout). Thus, dear readers, we end this chapter with a reprise of the classical test for apocalypses—the theme that has circulated throughout these pages, and throughout Western history. This book will bear a November 1997 publication date. But if the theory of 6000 *Annus Mundi* holds, and if Ussher got his chronology right, the world will end on October 23, 1997. So, if you are reading this book—as I fervently hope you are—then the anticipated apocalypse has been postponed once again. The only truly repeated pattern of the ages—the failure of apocalyptic predictions—has played one more time to perfuse our spirits with the satisfaction of a knowable world order. God must be in His heaven—and all must be right with the world!

2

When?

DOUSING DIMINUTIVE
DENNIS'S DEBATE
(DDDD = 2000)

In 1697, on the day appointed for repenting mistakes in judgment at Salem, Samuel Sewall of Boston stood silently in old South Church, as the rector read his confession of error aloud and to the public. He alone among judges of the falsely accused "witches" of Salem had the courage to undergo such public chastisement. Four years later, the same Samuel Sewall made a most joyful noise unto the Lord—and at a particularly auspicious moment. He hired four trumpeters to herald, as he wrote, the "entrance of the 18th century" by sounding a blast on Boston Common right at daybreak. He also paid the town crier to read out his "verses upon the New Century." The opening stanzas seem especially poignant today, the first for its relevance (I am writing this essay on a bleak January day in Boston, and the temperature outside is −2° Fahrenheit), and the second for a superannuated paternalism that highlights both the admirable and the dubious in our history:

Once more! Our God vouchsafe to shine:
Correct the coldness of our clime.
Make haste with thy impartial light,
and terminate this long dark night.
Give the Indians eyes to see
The light of life, and set them free.
So men shall God in Christ adore,
And worship idols vain, no more.

I do not raise this issue either to embarrass the good judge for his tragic error, or to praise his commendable courage, but for an aspect of the tale that may seem peripheral to Sewall's intent, yet nevertheless looms large as we approach the millennium destined to climax our current decade. Sewall hired his trumpeters for January 1, 1701, not January 1, 1700—and he therefore made an explicit decision in a debate that the cusp of his new century had kindled, and that has increased mightily at every similar transition since (see my main source for much of this section, the marvelously meticulous history of fins de siècle—*Century's End* by Hillel Schwartz). When do centuries end? At the termination of years marked '99 (as common sensibility suggests), or at the close of years marked '00 (as the narrow logic of a particular system dictates)?

The debate is already more intense than ever, though

Condemned in Hell (1499–1500), Luca Signorelli. Fresco.

we still have a little time before our own forthcoming transition, and for two obvious reasons. First—O cursèd spite—our disjointed times, and our burgeoning press, provide greatly enhanced opportunity for rehearsal of such narrishkeit ad nauseam; do we not feast upon trivialities to divert attention from the truly portentous issues that engulf us? Second, this time around really does count as the ultimate blockbuster: for this is the millennium,* the great and indubitable unicum of any living observer (though a few trees, and maybe a fungus or two, but not a single animal, were born before the year 1000 and have therefore been through it before).

On December 26, 1993, *The New York Times* ran a piece to bury the Christmas buying orgy and welcome the new year. This article, on commercial gear-up for the century's end, began by noting: "There is money to be made on the millennium . . . in 999 feelings of

---

*In this book's spirit of dispelling a standard set of confusions that have surrounded the forthcoming millennium, may I at least devote a footnote to the most trivial but also the most unambiguously resolvable. *Millennium* has two n's—honest to God, it really does, despite all the misspellings, even in most of the books and product names already dedicated to the event. The adjective *millennial* also has two, but the alternative *millenarian* only has one. (The etymologies are slightly different. *Millennium* is from the Latin *mille,* "one thousand," and *annus,* "year"—hence the two n's. *Millenarian* is from the Latin *millenarius,* "containing a thousand (of anything)," hence no *annus,* and no two n's.)

gloom ran rampant. What the doomsayers may have lacked was an instinct for mass marketing." The commercial cascade of this millennium is now in full swing: in journals, date books, the inevitable coffee mugs and T-shirts, and a thousand other products being flogged by the full gamut, from New Age fruitcakes of the counterculture, to hard-line apocalyptic visionaries at the Christian fringe, to a thicket of ordinary guys out to make a buck. The article even tells of a consulting firm explicitly established to help others market the millennium—so we are already witnessing the fractal recursion that might be called metaprofiteering, or growing clams of advice in the clam beds of your advisees' potential profits.

I am truly sorry that I cannot, in current parlance, "get with the program." I feel compelled to mention two tiny little difficulties that could act as dampers upon the universal ballyhoo. First, millennia are not transitions at the ends of thousand-year periods, but particular periods lasting one thousand years; so I'm not convinced that we even have the name right (but see Part 1 for a resolution of this issue). Second, if we insist on a celebration (as we should) no matter what name be given, we had better decide when to celebrate. I devote this section to explaining why the second issue cannot be resolved—a situation that should be viewed as enlightening, not depressing. For just as Tennyson taught us to prefer love lost over love unexperienced, it is better to not know and to know why one can't know, than to be clueless about why the hell so many people are so agitated about 1999 versus 2000 for the last year before the great divide. At least when you grasp the conflicting, legitimate, and unresolvable claims of both sides, you can then celebrate both alternatives with equanimity—or neither (with informed self-righteousness) if your persona be sour, or smug.

As a man of below average stature myself, I am delighted to report that the source of our infernal trouble about the ends of centuries may be traced to a sixth century monk named Dionysius Exiguus, or (literally) Dennis the Short. Instructed to prepare a chronology for Pope St. John I, Little Dennis, following a standard practice, began countable years with the foundation of Rome. But, neatly balancing his secular and sacred allegiances, Dionysius then divided time again at Christ's appearance. He reckoned Jesus' birth at December 25, near the end of year 753 A.U.C. (*ab urbe condita,* or "from the foundation of the city," that is, of Rome). Dionysius then restarted time just a few days later on January 1, 754 A.U.C.—not Christ's birth, but the feast of the circumcision on his eighth day of life, and also, not coincidentally, New Year's Day in Roman and Latin Christian calendars.

Dionysius's legacy has provided little but trouble. First of all, as discussed in more detail in Part 1, he didn't even get the date right, for Herod died in 750 A.U.C. Therefore, if Jesus and Herod overlapped (and the gospels will have to be drastically revised if they did not), then Jesus must have been born in 4 B.C. or ear- lier—thus granting the bearer of time's title several years of life before the inception of his own era!

(I do, in any case, relish the oxymoron of Jesus born at least four years before Jesus. For various reasons, including resolution of this paradox and a desire for greater inclusivity in a diverse world containing lots of non-Christian folks, the B.C. terminology has been los- ing popularity of late. Some sources now use B.C.E.—for "before the Christian era" if they wish to tone down the oxymoron, or "before the common era" if they care about inclusivity. Scientists, recognizing absolutely nothing special about the B.C.–A.D. transition, tend to use B.P., or "before the present," as in 32,410 B.P. for the oldest radiocarbon dated Paleolithic cave painting from Chauvet in France—a good way to acknowledge the anachronistic irrelevance of Jesus' birth for an ear- lier cave artist, but carrying the obvious disadvantage that the present gains an increment of one each year, and any B.P. date can only be interpreted if you know the year of publication for your source. God (and Jesus) forbid if you have lots of B.P. dates from publications of widely different years, and then have to correct them all to a common standard.)

But Dennis's misdate of Jesus counts as a mere peccadillo compared with the consequences of his second bad decision. He started time again on January 1, 754 A.U.C.—and he called this date January 1 of year one A.D. (*Anno Domini,* or "in the year of the Lord")—not year zero (which would, in retrospect, have spared us ever so much trouble!). In short, Dennis neglected to begin time at zero, thus discombobulating all our usual notions of counting. During the year that Jesus was one year old, the time system that supposedly started with his birth was two years old. (Babies are zero years old until their first birthday; modern time was already one year old at its inception.)

We should not, however, be overly harsh on poor Dennis—for this most inconvenient error (in retrospect) could not have been avoided, and surely cannot be laid on his doorstep (if monastical cubicles even included such an architectural feature for absorbing metaphorical blame). Western mathematics in the sixth century had not yet developed a concept of zero to serve as a proper place marker across Dennis's great divide. The Egyptians had used a zero, but only sporadically and inconsistently. The Chinese had no explicit numeral for zero, but their abacus implied the concept. The Mayans did develop a symbol for zero, but could not use the concept in a fully systematic way in their calculations (not to mention that Dennis knew absolutely nothing either of them or their entire hemisphere). Hindu and Arabic mathematicians devised the concept of zero in a complete and usable way—but not, apparently, before the late eighth or early ninth century— and Europe borrowed the idea from this source. Ironically, another figure in our narrative, the millennial Pope (and great scholar) Sylvester II, who reigned as pontiff from 999 to 1003, became the major exponent of zero, and our modern Arabic system of numbers—but far too late for Dennis (and for surcease from the perpetual confusion that has reigned ever since).

The problem of centuries arises from Dennis's unfortunate, if historically inevitable, decision to start at one, rather than zero—and for no other reason! If we insist that all decades must have ten years, and all centuries one hundred years, then year 10 belongs to the first decade—and, sad to say, year 100 must remain in the first century. Thenceforward, the issue never goes away.

Every year with a '00 must count as the hundredth and final year of its century—no matter what common sensibility might prefer: 1900 went with all 1800 years to form the nineteenth century; and 2000 must be the completing year of the twentieth century, not the inception of the next millennium. Or so the pure logic of Dennis's system dictates. If our shortsighted monk had only begun with a year zero, then logic and sensibility would coincide, and the wild millennial bells could ring forth but once and resoundingly at the beginning of January 1, 2000. But he didn't.

Since logic and sensibility do not coincide, and since both have legitimate claims upon our decision, the great and recurring debate about century boundaries simply cannot be resolved. Some questions have answers because obtainable information decrees a particular conclusion. The earth does revolve around the sun, and evolution does regulate the history of life. Some questions have no answers because we cannot get the required information. (I doubt that we will ever identify Jack the Ripper with certainty.) Many of our most intense debates, however, are not resolvable by information of any kind, but arise from conflicts in values or modes of analysis. (Shall we permit abortion, and in what circumstances? Does God exist?) A subset of these unresolvable debates—ultimately trivial, but

Detail from *Condemned in Hell* (1499–1500), Luca Signorelli.

capable of provoking great agitation, and thus the most frustrating of all—have no answers because they are about words and systems, rather than things. Phenomena of the world (that is, "things") therefore have no bearing upon potential solutions. The century debate lies within this vexatious category.

The logic of Dionysius's arbitrary system dictates one result—that centuries change between '00 and '01 years. Common sensibility leads us to the opposite conclusion: We want to match transitions with the extent or intensity of apparent sensual change, and 1999 to 2000 just looks more definitive than 2000 to 2001. So we set our millennial boundary at the change in all four positions, rather than the mere increment of 1 to the last position. (I refer to this side as "common sensibility" rather than "common sense" because support invokes issues of aesthetics and feeling, rather than logical reasoning.)

One might argue that humans, as creatures of reason, should be willing to subjugate sensibility to logic; but we are, just as much, creatures of feeling. And so the debate has progressed at every go-round. Hillel Schwartz, for example, cites two letters to newspapers, written from the camp of common sensibility in 1900: "I defy the most bigoted precisian to work up an enthusiasm over the year 1901, when we will already have had twelve months' experience of the 1900's." "The centurial figures are the symbol, and the only symbol, of the centuries. Once every hundred years there is a change in the symbol, and this great secular event is of startling prominence. What more natural than to bring the century into harmony with its only visible mark?" Since these strong expressions precede the invention of the automobile odometer, we cannot attribute current preferences for honoring 2000 to the most obvious device that now concentrates our attention upon the numerical side of millennial transitions. (My dad once took me and my brother on a late night ten-mile ride around Flushing—just so we could see the odometer go from 9999 to 10000—rather than giving him the pleasure on his solo trip to the office next morning. I'll bet that half the readers of this essay could cite a similar experience.)

I do so love human foibles; what else can keep us laughing (as we must) in this tough world of ours. The more trivial an issue, and the more unresolvable, so does the heat of debate, and the assurance of absolute righteousness, intensify on each side. (Just consider professorial arguments over parking places at university lots.) The same clamor arises every hundred years. An English participant in the debate of 1800 versus 1801 wrote of "the idle controversy, which has of late convulsed so many brains, respecting the commencement of the current century." On January 1, 1801, a poem in the Connecticut *Courant* pronounced a plague on both houses (but sided with Dionysius):

Precisely twelve o'clock last night,
The Eighteenth Century took its flight.
Full many a calculating head
Has rack'd its brain, its ink has shed,
To prove by metaphysics fine
A hundred means but ninety-nine;
While at their wisdom others wonder'd
But took one more to make a hundred.

The same smugness reappeared a century later. *The New York Times,* with anticipatory diplomacy, wrote in 1896: "As the present century draws to its close we see looming not very far ahead the venerable dispute which reappears every hundred years—*viz:* When does the next century begin? . . . There can be no doubt that one person may hold that the next century begins on the 1st of January, 1900, and another that it begins on the 1st of January, 1901, and yet both of them be in full possession of their faculties." But a German commentator remarked: "In my life I have seen many people do battle over many things, but over few things with such fanaticism as over the academic question of when the century would end. . . . Each of the two parties produced for its side the trickiest of calculations and maintained at the same time that it was the simplest matter in the world, one that any child should understand."

You ask where I stand? Well, publicly of course I take no position because, as I have just stated, the issue is unresolvable: for each side has a fully consistent argument within the confines of different but equally defensible systems. But privately, just between you and me, well, let's put it this way: I know a mentally handicapped young man who also happens to be a prodigy in day-date calculation. (He can, instantaneously, give the day of the week for any date, thousands of years, past or future—see Part 3.) He is fully aware of the great century debate, for nothing could interest him more. I asked him recently whether the millennium comes in 2000 or 2001—and he responded unhesitatingly: "In 2000. The first decade had only nine years."

What an elegant solution, and why not? After all, no one then living had any idea whether they were toiling in year zero or year one—or whether their first decade had nine or ten years, their first century ninety-nine or one hundred. The B.C.–A.D. system wasn't invented until the sixth century and wasn't generally accepted in Europe until much later. So why don't we just proclaim that the first century had ninety-nine years—since not a soul then living either knew or cared about the

Detail from *The Last Judgement* (1536–1541), Michelangelo.

anachronism that would later be heaped upon all the years of their lives? Centuries can then turn when common sensibility desires, and we underscore Dionysius's blessed arbitrariness with a caprice, a device of our own that marries the warring camps. Neat, except that I think people want to argue passionately about trivial unresolvabilities—lest they be compelled to invest such rambunctious energy in real battles that might kill somebody.

What else might we salvage from rehearsing the history of a debate without an answer? Ironically, such arguments contain the possibility for a precious sociological insight: Since no answer can arise from either the factuality of nature or the internal necessities of human logic, changing viewpoints provide "pure" trajectories of evolving human attitudes—and we can therefore map societal trends without impediments of such confusing factors as definite truth.

I had intended to spend only a few hours in research for this chapter, but as I looked up documents from century transitions, I noticed something interesting in this sociological realm. The two positions—I have called them "logical" and "common sensible" so far in this chapter—also have clear social correlations that I had not anticipated. The logical position—that centuries must have one hundred years and transitions must therefore occur, because Dionysius started at one rather than zero, between '00 and '01 years—has always been overwhelmingly favored by scholars and by people in power (the press and business in particular), representing what we may call "high culture." The common sensible position—that we must honor the appearance of maximal change between '99 and '00 years and not fret overly about Dionysius's unfortunate lack of foresight—has been the perpetual favorite of that mythical composite once designated as John Q. Public, or the "man in the street," and now usually called vernacular or pop culture.

The distinction goes back to the very beginning of this perpetually recurring debate about century transitions. Hillel Schwartz traces the first major hassle to the 1699–1701 passage (place the moment where you wish), the incarnation that prompted Samuel Sewall's trumpeting in Boston. Interestingly, part of the discussion then focused upon an issue that has been persistently vexatious ever since: *viz.*, did the first millennial transition of 999–1001 induce a period of fear about imminent apocalyptical endings of the world?

I discussed this topic in Part 1 and wish now only to point out that the first published claim for a panic terror, a late sixteenth-century work by Cardinal Cesare Baronio, also addressed the great issue of endings for centuries—as this document of undoubted high culture favored the end of the year 1000 for apocalyptic expectations, while most popular writing has always focused on the end of 999 (as in the newspaper quotation cited on page 104). Thus, whether by anachronism or direct testimony, this debate has always been with us. Hillel Schwartz writes:

> Sarcastic, bitter, sometimes passionate debates in re a terminus on New Year's Eve '99 vis-à-vis New Year's '00, have been prosecuted since the 1690's and confusion has spread to the mathematics of the millennial year. For Baronio and his (sparse) medieval sources, the excitements of the millennium were centered upon the end of the year 1000, while the end of 999 has figured more prominently in the legend of the panic terror.

The pattern has held ever since, as the debate bloomed in the 1690s, spread in the 1790s with major centers in newspapers of Philadelphia and London (and with added poignancy as America mourned the death of George Washington at the very end of 1799), and burst out all over the world in a frenzy of discussion during the 1890s.

The 1890s version displays the clearest division of

high versus vernacular culture. A few high culture sources did line up behind the pop favorite of 1899–1900. Kaiser Wilhelm II of Germany officially stated that the twentieth century had begun on January 1, 1900. A few barons of scholarship, including such unlikely bedfellows as Sigmund Freud and Lord Kelvin, agreed. But high culture overwhelmingly preferred the Dionysian imperative of 1900–1901. An assiduous survey showed that the presidents of Harvard, Yale, Princeton, Cornell, Columbia, Dartmouth, Brown, and the University of Pennsylvania all favored 1900–1901—and with the entire Ivy League so firmly behind Dionysius, why worry about a mere Kaiser?

In any case, 1900–1901 won decisively, in the two forums that really matter. Virtually every important public celebration for the new century, throughout the world (and even in Germany), occurred from December 31, 1900, into January 1, 1901. Moreover, essentially every major newspaper and magazine officially welcomed the new century with their first issue of January 1901. I made a survey of principal sources and could find no exceptions. *The Nineteenth Century,* a leading British periodical, changed its name to *The Nineteenth Century and After,* but only with the January 1901 issue, which also featured a new logo of bifaced Janus, with an old bearded man looking down and left into the nine-

THE

NINETEENTH

CENTURY

AND AFTER

XIX-  -XX

A MONTHLY REVIEW

EDITED BY JAMES KNOWLES

VOL. XLIX

JANUARY—JUNE 1901

LONDON
SAMPSON LOW, MARSTON & COMPANY
(LIMITED)
St. Dunstan's House
FETTER LANE, FLEET STREET, E.C.
1901

The Nineteenth Century and After: A Monthly Review (1901).

Entered at the New-York Postoffice as Second-class Matter.

LIBRARY OF TRIBUNE EXTRAS.

VOL. XIII. **JANUARY, 1901.** **NO. 1.**

TRIBUNE ALMANAC

19 01

AND

Political Register

FIRST NUMBER OF

The Twentieth Century

25 Cents a Copy; $2.00 a Year.

Tribune Almanac and Political Register (1901).

teenth century and a bright youth looking right up into the twentieth. Such reliable standards as *The Farmer's Almanack* and *The Tribune Almanac* declared their volumes for 1901 as "first number of the twentieth century." On December 31, 1899, *The New York Times* began a story on the nineteenth century by noting: "Tomorrow we enter upon the last year of a century that is marked by greater progress in all that pertains to the material well-being and enlightenment of mankind than all the previous history of the race." A year and a day later, on January 1, 1901, the lead headline proclaimed "Twentieth Century's Triumphant Entry" and described the festivities in New York City: "The lights flashed, the crowds sang, the sirens of craft in the harbor screeched and roared, bells pealed, bombs thundered, rockets blasted skyward, and the new century made its triumphant entry." Meanwhile, poor Carry Nation never got to watch the fireworks, or even to raise a glass, for a small story on the same first page announced: "Mrs. Nation Quarantined—Smallpox in jail where Kansas saloon wrecker is held—says she can stand it."

Thus, the last time around, high culture still held the reins of opinion—even in such organs of pop culture as *The Farmer's Almanack,* no doubt published by men who considered themselves among the elite. But consider the difference as we approach the millennium—for who can doubt that pop culture will win decisively on this most important of all replays? Oh, to be sure, the "official" sources of a waning purity in high culture will make their customary noises. Indeed, as I was revising this essay, I noted the following headline in *The New York Times* for December 8, 1996: "British Observatory Takes Stand on When Millennium Begins." The story begins by acknowledging the fait accompli of pop culture's imposition this time around:

> When the clock strikes midnight on December 31, 1999, billions of people around the world will celebrate the dawn of a new millennium—a year too early, some experts say. As the champagne flows and kisses mark the start of the new age, the revelers will actually be welcoming the last year of the present millennium, not the first year of the next, they say.

The *Times* then reports that the most official of all conceivable sources—the gold standard that could easily have imposed its will in centuries past—has thrown down the gauntlet for high culture's perennial favorite, Dionysius Exiguus's unpopular solution: "The start of the new millennium is January 1, 2001—not the year 2000, say researchers at the Royal Greenwich Observatory in Cambridge, England."

Times have changed, however, and the *Times* quickly acknowledged why high culture's Greenwich solution cannot prevail. First of all, no one now wields an imprimatur in our decentralized world:

> In addition to no longer being in Greenwich, the observatory is no longer the world's timekeeper. "Coordinated universal time" measured by some 150 atomic clocks around the world has replaced Greenwich mean time as the standard.

Second, pop culture's preferences can no longer be denied. Even once mighty Greenwich has been reduced to impotent tut-tutting! The *Times* story continues:

> The year 2000 "will certainly be celebrated, as is natural for a year with such a round number," a statement issued by the observatory said. "But, accurately speaking," it said, "we will be celebrating the 2,000th year, or the last year of the millennium, not the start of the new millennium."

True to form, but armed this time with the invincible authority of new social relations, pop culture will have none of John Bull's bushwa. Take cover: the perennial (or rather percenturial) debate is on once again! Two letters appeared in the *Times* on December 12, one announcing with bored insouciance that it's all over anyway because more than two thousand years have elapsed since Christ's actual birth; the other responding with scorn and vigor to the old guard of Greenwich:

> Enough already with the sophistic explanations of why the year 2000 is not the beginning of the new millennium. Popular wisdom will make it so, even if the astronomers disbelieve. Their argument that there is no year zero is silly; we can have a year zero any time we want. The sequence of years can be redefined as 3 B.C., 2 B.C., 1 B.C. *or* 0 A.D., 1 A.D., 2 A.D. . . . Then the year 1 B.C. would merely have different names in the A.D. system and the B.C. system.

This letter provides yet another clever, and perfectly adequate, rationale for celebrating in 2000—a lovely solution akin to my informant's conviction (cited previously) that the first century had only ninety-nine years. As I have emphasized throughout, arbitrary problems without conceivable final answers require consistent but arbitrary solutions.

In any case, and in the truly decisive court of culture and sociology, who can doubt that 2000 will win this time? Arthur C. Clarke and Stanley Kubrick stood by Dionysius in book and film versions of *2001*, but I can hardly think of another source that does not specify the inception of 2000 as the great moment of transition. All book titles of our burgeoning literature honor pop culture's version of maximal numerical shift—including Ben Bova's *Millennium: A Novel about People and Politics in the Year 1999;* J. G. de Beus's *Shall We Make the Year 2000;* Raymond Williams's *The Year 2000;* and even Richard Nixon's *1999: Victory Without War.* Prince's album and lead song *1999* cites the same date from this ne plus ultra of pop sources.

Cultural historians have often remarked that the expansion of pop culture, including both respect for its ways and diffusion of its influence, marks a major trend of the twentieth century. Musicians from Benny Goodman to Wynton Marsalis play their instruments in jazz bands and classical orchestras. The Metropolitan Opera has finally performed *Porgy and Bess*—and bravo for them. Scholars write the most damnedly learned articles about Mickey Mouse.

This remarkable change has been well documented and much discussed, but commentary has so far missed this important example from the great century debate. The distinction still mattered in 1900, and high culture won decisively by imposing January 1, 1901 as the inception of the twentieth century. Pop culture (or the amalgam of its diffusion into courts of decision makers) may already declare clear victory for the millennium, which will occur at the beginning of the year 2000 because most people so feel it in their bones, Dionysius notwithstanding—and again I say bravo. My young friend wanted to resolve the debate by granting the first century only ninety-nine years; now ordinary humanity has spoken for the other end—and the transition from high culture dominance to pop culture diffusion will resolve this issue of the ages by granting the twentieth century but ninety-nine years! The old guard of Greenwich may pout to their heart's content, but the world will rock and party on January 1, 2000.

How lovely—for eternal debates about the unresolvable really do waste a great deal of time, put us in bad humor, and sap our energy from truly important pursuits. Let us, instead, save our mental fight—not to establish the blessed millennium (for I doubt that humans are capable of such perfection) but at least to build Jerusalem upon our planet's green and pleasant land.

3

Why?

PART ONE:
BLOODY-MINDED NATURE

We have a false impression, buttressed by some famously exaggerated testimony, that the universe runs with the regularity of an ideal clock, and that God must therefore be a consummate mathematician. In his most famous aphorism, Galileo described the cosmos as "a grand book written in the language of mathematics, and its characters are triangles, circles, and other geometrical figures." The Scottish biologist D'Arcy Thompson, one of my earliest intellectual heroes and author of the incomparably well-written *Growth and Form* (first published 1917 and still vigorously in print, the latest edition with a preface by yours truly), stated that "the harmony of the world is made manifest in Form and Number, and the heart and soul and all the poetry of Natural Philosophy are embodied in the concept of mathematical beauty."

Many scientists have invoked this mathematical regularity to argue, speaking metaphorically at least, that any creating God must be a mathematician of the

The Vision of Saint John, El Greco (1541–1614).

Pythagorean school. For example, the celebrated physicist James Jeans wrote: "From the intrinsic evidence of his creation, the Great Architect of the Universe now begins to appear as a pure mathematician." This impression has also seeped into popular thought and artistic proclamation. In a lecture delivered in 1930, James Joyce defined the universe as "pure thought, the thought of what, for want of a better term, we must describe as a mathematical thinker."

If these paeans and effusions were invariably true, I could compose my own lyrical version of the consensus, and end this book forthwith. For I have arrived at the last great domain for millennial questions—calendrics. I need to ask why calendrical issues have so fascinated people throughout the ages, and why so many scholars and mathematicians have spent so much time devising calendars and engaging in endless debates about proper versus improper, elegantly simple versus overly elaborate, natural versus contrived systems for counting seconds, minutes, hours, days, weeks, months, lunations, years, decades, centuries and millennia, tuns and baktuns, tithis and karanas, ides and nones. Our culturally contingent decision to recognize millennia, and to impose divisions by 1,000 upon a solar system that includes no such natural cycle, adds an important ingredient to this maelstrom of calendrical debate.

If God were Pythagoras in Galileo's universe, calendrics would never have become an intellectual subject at all. The relevant cycles for natural timekeeping would all be nice, crisp, easy multiples of each other—and any fool could simply count. We might have a year (earth around sun) with exactly ten months (moon around earth), and with precisely one hundred days (earth around itself) to the umpteenth and ultimate decimal point of conceivable rigor in measurement. But God, thank goodness, includes both Loki and Odin, the comedian and the scholar, the jester and the saint. God did not fashion a very regular universe after all. And we poor sods of his image are therefore condemned to struggle with calendrical questions till the cows come home, and Christ comes round again to inaugurate the millennium.

Oh, I don't deny that some corners of truly stunning mathematical regularity grace the cosmos in domains both large and small. The cells of a honeybee's hive, and the basalt pillars of the Giant's Causeway in Northern Ireland, make pretty fair and regular hexagons. Many "laws" of nature can be written in an astonishingly simple and elegant mathematical form. Who would have thought that $E = mc^2$ could describe the unleashing of the prodigious energy in an atom?

But we have been oversold on nature's mathematical

regularity—and my opening quotations stand among the worst offenders. If anything, nature is infinitely diverse and constantly surprising—in J. B. S. Haldane's famous words, "not only queerer than we suppose, but queerer than we *can* suppose." I call this section "Bloody-Minded Nature" because I wish to specify the two opposite domains of nature's abject refusal to be mathematically simple for meaningful reasons. The second domain forces every complex society—as all have independently done, from Egypt to China to Mesoamerica—to struggle with calendar-making as a difficult and confusing subject, not a simple matter of counting. Many questions about the millennium—Why do we base calendars on cycles at all? Why do we recognize a thousand-year interval with no tie to any natural cycle?—arise directly from these imposed complexities. Any adequate account of our current millennial madness therefore requires that we understand why calendrics has been such a troubling and fascinating subject for all complex human societies.

In the first domain, apparent regularities turn out to be accidental—and the joke is on us. In the most prominent example, consider the significance and importance that traditional culture invested in the equal size of the sun and moon in the sky—a major source of richness for our myths and sagas, and a primary ingredient in our recipe for meaningful order in the heavens: "And God made two great lights: the greater light to rule the day, and the lesser light to rule the night" (Genesis 1:16). But the equality in observed size is entirely fortuitous, and not a consequence of any mathematical regularity or law of nature. The sun's diameter is about four hundred times larger than the moon's, but the sun is also about four hundred times more distant—so the two discs appear nearly identical in size to an observer on earth.

In the second and opposite domain, deeply useful and earnestly sought regularities simply do not exist—and we must resort to inconvenient approximations and irreducible unevenness. The complexities of calendrics arise almost entirely within this domain—and I shall illustrate this essential point with the two primary examples that have dogged humanity ever since Og the Caveperson first recognized that his full-moon symbols, all neatly and carefully inscribed on his mammoth-shoulderblade scratching board, did not line up evenly with the day symbols carved into the row just below. So Og scratched his head, decided that he must have made a mistake, kept his records even more carefully, and always got the same uneven result. (Og either went mad, became a crashing bore to his fellows and ended up in exile, or went with the empirical flow and became the first architect of a complex and approximate calendar.)

The two primary examples that have plagued all complex cultures—the fractional number of days and lunations in the solar year—arise from the same source: nature's stubborn refusal to work by simple numerical relations in the very domain where such regularity would be most useful to us. Nature, apparently, can make a gorgeous hexagon, but she cannot (or did not deign to) make a year with a nice even number of days or lunations.

What a bummer. Both our practical requirements (to know the seasons for hunting or agriculture, and the tides for fishing or navigation, not to mention that great bugaboo of Christian history, the calculation of Easter), and our intrinsic mental need to seek numerical regularity as one way of ordering a confusing world, drive us to keep track of the three great natural cycles—the days of the earth's rotation, the lunations of the moon's revolution, and the years of the earth's revolution. (Our other major cycles, from weeks to millennia, do not map astronomical events, and arise for more complex and contingent reasons of human history.)

If any of these three natural cycles worked as an even multiple of any other, we could have such a nice, easy, and recurrent calendar—making life ever so much more convenient. Nature, however, gives us nothing but fractionality to innumerable and nonending decimal places—and so it goes. We may best gauge how this inconvenient construction of reality has affected human history by tracing how Western society has treated the two great calendrical complexities imposed by nature's noncoincident cycles.

The Days of the Solar Year

365 days, 5 hours, 48 minutes, and 45.96768 . . . seconds! What hath God wrought? The Egyptians found out, so did the Chinese, and so did the Mayans—all independently, and all to their dismay. 365.25—exactly 365 and an extra quarter day precisely—would have been bad enough. We would still face the inconvenience of a leap year every four years, with all the attendant lore—including a variable February that gobbles up a full two-thirds of the six-line ditty that once taught every schoolchild the lengths of the months:

> *Thirty days hath September*
> *April, June and November.*
> *All the rest have thirty-one,*
> *Except for February alone*
> *Which hath but twenty-eight, in fine,*
> *Till leap year grants it twenty-nine.*

The Last Judgement, William Blake.

The power of such doggerel can be daunting. To this day, I cannot separate the 30s from the 31s without intoning the first two lines in their entirety.

More rational solutions can easily be devised to regularize the intermediary units that many calendars utilize and that we call months, even though they run out-of-whack with true lunations (for good reasons discussed in the next section). Several societies independently hit upon the idea of dividing 360 days into equal divisions (18 "months" of 20 days each for the vigesimal Mayans; 12 newly named months of 30 days each for the revolutionary French in their wipe-the-slate-clean-and-start-again calendar of 1792)—and then proclaiming five special days to round out the year (viewed as especially unlucky by the Mayans, but as a grand excuse for a long party by the French). Fair enough, but you still have to deal with that pesky extra quarter day each year. So the French added an extra special day, for six *in toto,* every four years.

The riddles of leap year can provoke endless complexity and wondrously trivial discussion. Just consider all the birthday lore, and the tales of great characters both actual and fictional. Take the case of the ever-youthful composer Rossini, who recently celebrated his forty-eighth birthday on February 29, 1992, just after the earth completed its two hundredth circuit around the sun since his birth in 1792. (Yes, his forty-eighth, not his fiftieth birthday; hang on a bit, for this vexatious little item requires the next level of calendrical complexity, discussed in the next section, for a resolution.)

And consider the poor pirate 'prentice Frederick, indentured to the notorious Pirates of Penzance until his twenty-first *birthday*. Gilbert and Sullivan's comic opera of the same name begins with celebrations for Frederick's forthcoming release. But the poor lad was born on February 29, so he is only "five and a little bit over." The opera bears the subtitle "The Slave of Duty"—so you can figure out that Frederick agrees to stay until the contractually appointed time. He then importunes his fiancée Mabel: "In 1940 I of age shall be; I'll then return and claim you, I declare it." Mabel replies, "It seems so long," but then promises to wait. Poor Mabel. The situation is bad enough already, but Gilbert—as we shall see—made the same mistake as folks who thought that Rossini had celebrated his fiftieth birthday in 1992. Mabel must really wait until 1944, when Frederick will be a spry eighty-eight, not 1940, at her beau's distinctly more youthful eighty-four chronological years.

The first modern reform of the Western calendar, introduced by Julius Caesar himself in 45 B.C., didn't recognize the additional irregularity of 365-and-a-teeny-little-bit-less-than-a-quarter-of-a-day (365.242199 . . . to be precise) and used exactly 365-and-a-quarter instead. Can we possibly need to worry about such a minor rounding-off that overestimates the true solar year by a mere eleven minutes and change? Thus, the Julian calendar operated in a maximally simple manner (given the undeniable reality of that fractional day after the full 365). That is, the Julian calendar makes one correction, and one correction only—and this correction follows an invariable rule. On every fourth or "leap" year, the calendar adds an extra day to make a year of 366 days. Since we cannot abide fractional days in a rational calendar, an endlessly repeating sequence of 365, 365, 365, and 366 will serve as a good whole-day version of a solar year that actually runs for 365 and a quarter days.

Except for the inconvenient additional complexity that the solar year doesn't quite reach 365 and a quarter days. The year falls short of this fractional regularity by those pesky eleven minutes and change. The minor overestimate of the Julian calendar will not matter much at first, but those eleven extra minutes do begin to add up after a while, and Caesar did live a rather long time ago. Eventually, the calendar will start to accumulate noticeable extra days (seven every thousand years, in fact), and the process must continue indefinitely, forcing the Julian calendar more and more out of whack with the true solar year. That is, if we want the vernal equinox to fall on about the same day, March 21 or so, every year (an enormous convenience for all manner of people, from priests to farmers, and a pressing necessity, as we shall see, for the crucial determination of Easter), then the Julian calendar gets progressively worse as the centuries roll. The vernal equinox (and any other fixed date) begins to creep farther and farther up the calendar. And this blot on Caesar's reputation, rather than Brutus's wound, may turn out to be the most unkindest cut of all.

Pope Gregory XIII therefore made a kind and rational cut instead. By the sixteenth century, this inexorable overestimate, ticking along at eleven minutes and fourteen seconds per year, had accumulated ten extra days. This sloppiness had begun to generate some serious consequences, particularly for priests and astronomers charged with the solemn and sacred duty of determining the date for Easter. So Gregory followed a strategy favored from time immemorial—he convened a committee and appointed a very smart chairman, the eminent Jesuit mathematician Christopher Clavius. This committee, beginning its work in 1578, came up with one of those lovely, practical solutions that has absolutely no mandate in elegant or highfalutin theory but possesses the cardinal virtue of working pretty damned well. Pope Gregory proclaimed the new rules in a papal bull issued on February 24, 1582. We call his correction the Gregorian reform, and the improved calendar—the one we still use today—the Gregorian calendar.

Clavius's committee faced two separate problems and solved them in different ways. First of all, the old Julian calendar was now running ten days ahead and had to be brought back into alignment with the solar year (so that equinoxes and solstices would fall at their traditional times—and stay put). This problem could only be solved by old fashioned damage control—of a fairly radical sort, but what else could they do? Clavius recommended that ten days be dropped into oblivion by official proclamation, and Pope Gregory did so—just like that, and by fiat! In 1582, October 5 through 14 simply disappeared and never occurred at all! The date following October 4 became October 15, and the calendar came back into sync.

This solution strikes many people as bizarre, if not monstrous—an affront both to nature and to human dignity. How can any arbitrary earthly power make days disappear at a whim? Now I don't deny that Gregory's solution imposed some problems (salaries, bank interest—if such a concept existed—ages, birthdays, and so on), though probably nothing on the scale of the forth-

Los, William Blake, from *The Book of Urizen*.

coming debacle (which we will try to prevent at great expense) when computers, on January 1, 2000, read their two-digit 00 year code as 1900 rather than 2000, and promptly go berserk with confusion. (I am, at least, looking forward to a hefty check for interest on an account that my bank has just read as on deposit for a hundred years.)

In fact, Gregory's solution of dropping days was not monstrous in the slightest but eminently wise and practical. The day records a true astronomical cycle, but the date that we affix to each day is only a human convention. October 5–14 were always part of an invented human system, not a natural reality. If we need to excise these dates in order to bring our artificial system into conformity with a natural cycle of equinoxes and solstices, then we may do so at will, and without guilt.

Secondly, Clavius and company had to devise a new calendrical rule that would avoid the creeping inaccuracy of the Julian system. They accomplished this goal by devising a year of 365.2422 days, much closer to astronomical reality than the calculationally simpler Julian solution of 365.25. To institute this new year, they made a second-order correction to the old rule of leap years—thus setting a more complex rule that we still use today. The Julian calendar had included too many leap years, so Clavius devised a neat little way to drop an occasional leap year in a regular manner that would give the entire system an appearance of wisdom and principle (thus hiding the purely practical problem that only required a workable and arbitrary solution). Clavius suggested that we drop the leap year at century boundaries, every hundred years.

But, as I argued at the outset of this section, natural cycles impose a numerical muddle—the very opposite of the adamantine order that Galileo or Jeans or Joyce wished to attribute to the cosmos. Simple rules rarely work, and the decision to drop leap years at century boundaries required yet another correction—third-order this time, with the Julian leap year as a first-order correction for the fractionality of days, the century drop as a second-order correction for the Julian overestimate, and this final rule as a third-order correction to the century drop.

Clavius recognized that if the Julian solution added too much, the century-dropping correction took away too much—requiring that something be put back every once in a while. Clavius therefore suggested that the leap year be restored every fourth century. He then expressed this procedure as a rule: Remove leap years at century boundaries, but put them back at century boundaries divisible by 400. (As I said, this may sound like a principled decision, but really represents no more than a codified rule of thumb.)

This third-order correction isn't perfect either, but it does bring the Gregorian calendar—that is, our calendar—into pretty fair sync with the solar year. In fact, the Gregorian year now departs from the solar year by only 25.96 seconds—accurate enough to require a correction of one day only once every 2,800 years or so. Finally, the discrepancy has become small enough not to matter in any practical way. (Or will these become famous last words as our technological society becomes ever more needful of precision?)

In summary, the Gregorian reform of 1582 revised the Julian calendar by dropping those "extra" ten days, and then promulgating a new rule of leap years to prevent any substantial future inaccuracy: Proclaim a leap year every four years, except for three out of four century boundaries; institute this rule by retaining the leap year at century boundaries divisible by 400. This Gregorian rule has an interesting consequence for the forthcoming millennial year 2000. What a special time, and what a privilege for all of us! Not only do we get to witness a millennial transition, but we also get to live in that rare year that comes only once in four hundred—a century boundary with a February 29. Yes, 2000 will be a leap year—and our lives will include the special bonus of an extra day that comes only once every four hundred years. Use it in good health!

As a final footnote to the subject of Gregorian corrections, this century rule explains the paradox of Rossini being only forty-eight years old after two hundred chronological years (1800 and 1900 were not leap years, so he didn't have a birthday), and poor Mabel's additional four years of waiting for Frederick to come of age (1900 was not a leap year).

So much for astronomy, but we also have to deal with the foibles of human history and human xenophobia. The truly improved Gregorian calendar was quickly accepted throughout the Roman Catholic world. But in England, the whole brouhaha sounded like a Popish plot, and the Brits would be damned if they would go along. Thus, England kept the Julian calendar until 1752, when they finally succumbed to reason and practicality—by which time yet another "extra" day had accumulated in the Julian reckoning, so Parliament had to drop eleven days (September 3–13, 1752) in order to institute the belated Gregorian reform.

When you know this history, some puzzling little footnotes in our common chronology gain an easy explanation (trivial in one sense, to be sure, but ever so frus-tratingly annoying if you don't know the reason). George Washington's birthday, for example, is some-times given, particularly by contemporary sources in colonial America, as February 11, 1731—rather than the February 22, 1732, that we used to celebrate on time, before all our public holidays moved to conve-nient Mondays and we decided to split the difference between Lincoln and Washington with a common Pres-idents' Day. As an English colony, America still used the Julian calendar at Washington's birth. The eleven days had not yet been dropped (so Gregorian February 22 still counted as Julian February 11 in the British world). Moreover, the Julian year began in March (at least in England), so Washington was born a year early as well.

Similarly, many people used to puzzle every year at the Soviet celebration of the "October Revolution" in November. (Remember all those tanks parading through Red Square past the Politburo on the bal-cony?) Russia did not adopt the Gregorian calendar until 1918, when the secularists ousted the orthodoxy. So the Julian October revolution had actually occurred in Gregorian November—those extra days again! Finally, since the enemy within is always more danger-ous than the enemy without, the Eastern Orthodox church has still not accepted the Gregorian calendar—

a Romish plot, no doubt. The Julian calendar still lives, but "there is a tide in the affairs of men . . ."

The Inconvenient Noncoincidence of the
Lunar and Solar Year

The day and the solar year fail to come into sync only by that tiny little bit less than a quarter of a day—but look at all the trouble so caused! When we turn to the moon, the situation deteriorates and, in fact, could hardly be worse.

The moon takes 29 and a half days to circle the earth (29.53059 days, to be more precise)—giving the natural month a horrendous fractionality when counted in days or factored into years. No regular "year" of lunations can therefore come even close to the solar year—the nearest approximation being twelve lunations of 354 total days (354.36706, to be more precise again), falling short of the solar year by almost eleven days.

This discrepancy might not matter if complex societies did not need to reconcile the lunar and solar years. But unfortunately though inevitably, they do—for the two reasons that have circulated throughout this text. First of all, practicality demands (for solar and lunar cycles are both so eminently, and differently, useful); secondly, reason also delights. (We are, for better or for worse, conscious creatures who wonder about our surroundings; we can scarcely observe the moon in its cycling phases and *not* ponder their regularity and their correlation with the other great cycles of days and years. We really have no choice.)

Many major societies in human history, notably imperial China, Judaism, and Islam, use a predominantly lunar calendar but also must establish reconciliation with the solar year. How can this be done? First of all, lunar months can't have fractional days, so you solve the problem of 29 and a half days per month by granting some of the twelve months 29 days (called "defective" or "hollow") and giving others (called "full") 30 days—all to make a full lunar year of 354 days. But what can be done about the eleven-day shortfall?

All societies with lunar calendars have struggled with this question, and all have discovered some variation of the so-called Metonic Cycle—another of those rough-and-ready rules of thumb disguised to look more like a principled law than a practical solution. When operating on such a coarse scale of 29- or 30-day lunar months, the simplest correction for the eleven-day shortfall in a year of lunations just adds an extra lunation—a "leap month," if you will—to make an occasional long year of thirteen lunar months or 384 days, whenever the accumulating shortfalls become troubling.

The Metonic Cycle, named for the fifth century B.C. Athenian astronomer Meton (but discovered earlier and independently in China and then again by Babylon, and thence into the Jewish calendar), recognizes the shortest sequence of years that can bring the solar and lunar calendars into nearly perfect alignment. The Metonic Cycle runs for nineteen years and requires the insertion of a leap month in any seven of those nineteen. (Actually, Meton's original version still included a discrepancy of five days after nineteen years, but this minor problem could easily be corrected by a variety of ad hoc solutions, including the addition of an extra day to five of those leap months.) The Metonic Cycle may sound rough and arbitrary; but it works, and nothing simpler can be devised. Therefore, nearly all lunar calendars follow this system of inserting leap months into seven of every nineteen years. The modern Jewish calendar, for example, intercalates a thirteenth month of 30 days in the third, sixth, eight, eleventh, fourteenth, seventeenth, and nineteenth years of a Metonic Cycle.

As with the Gregorian reform and the birth of George Washington, understanding such calendrical complexities really does help us to grasp some puzzling aspects of everyday life that otherwise persist as annoying confusions among the hundreds of little daily bothers that provoke the conventional response of a harried life: "Why the hell does it work this way? Someday I just have to look it up and find out"—and then we never do.

For example, didn't you always wonder why Chanukah creeps backward through the December calendar by about ten days every year? Then, just when you thought that Chanukah would sneak into November, it suddenly shoots forward the next year into late December, falling even after Christmas. Blame the Metonic Cycle. With respect to a Gregorian date, any Jewish date must move backward in any short lunar year of 354 days (twelve of nineteen in the Metonic Cycle), but shoot forward in long years of 384 days that add a leap month.

On the other hand, you may have noticed that Ramadan just keeps moving backward on our Gregorian calendar, and therefore can occur at any time during the solar year. The Islamic calendar is also lunar, but does not use the correction of the Metonic Cycle. The shortfalls therefore accumulate continually, and all fixed Islamic dates move constantly back on the Gregorian calendar.

In Christian history, the need to reconcile solar and lunar cycles has centered on one of the most complex and persistently vexatious problems in the history of

Guernica (1937), Pablo Picasso.

calendrics: the calculation of Easter. Books, indeed libraries, have been written on the subject, and great scholars have devoted their lives to devising rules and procedures for getting this cardinal day right. I shall not even begin to probe the details here but only wish to state, as a good summary for this part (for Easter, in its nutshell, epitomizes all the issues here discussed), that Easter became more problematic than any other calendrical day, or any other movable feast, because its definition includes both lunar and solar elements, and its date cannot be determined until we know how to reconcile all the great, and distressingly fractional, cosmic cycles. For Easter falls on the Sunday following the first full moon (the lunar component) after the vernal equinox (the solar contribution).

An Epilogue

I have always and dearly loved calendrical questions because they display all our foibles in revealing miniature. Where else can we note, so vividly revealed, such an intimate combination of all the tricks that recalcitrant nature plays upon us, linked with all the fallacies of reason, and all the impediments of habit and emotion, that make the fulfillment of our urge to understand even more difficult—in other words, of both the external and the internal pitfalls to knowledge. Yet we press on regardless—and we do manage to get somewhere.

I think that I love humanity all the more—the scholar's hangup, I suppose—when our urge to know transcends mere practical advantage. Societies that both fish and farm need to reconcile the incommensurate cycles of years and lunations. Since nature permits no clean and crisp correlation, people had to devise the cumbersome, baroque Metonic Cycle. And this achievement by several independent societies can only be called heroic.

I recognize this functional need to know, and I surely honor it as a driving force in human history. But when Paleolithic Og looked out of his cave and up at the heavens—and asked why the moon had phases, not because he could use the information to boost his success in gathering shellfish at the nearby shore, but because he just wanted to resolve a mystery, and because he sensed, however dimly, that something we might call recurrent order, and regard as beautiful for this reason alone, must lie behind the overt pattern—well, then calendrical questions became sublime, and so did humanity as well.

If we regard millennial passion in particular, and calendrical fascination in general, as driven by the pleasure of ordering and the joy of understanding, then this strange little subject—so often regarded as the province of drones or eccentrics, but surely not of grand or expansive thinkers—becomes a wonderful microcosm for everything that makes human beings so distinctive, so potentially noble, and often so actually funny. Socrates and Charlie Chaplin reached equal heights of sublimity.

I hate to end with such waffling generality, with such a risibly inadequate stab at lyricism—so let me finish instead with a little story about an ordinary person who has done something heroic in the domain of calendrics, and who loves the millennium with all his heart. His tale belongs to a classic genre in the annals of calendrics—*day-date calculation,* a subject that cannot be equaled (hence its classic status) as an illustration of interaction between human foibles and divine failure: that is, as something made difficult both because *we* chose peculiar definitions rather than eminently available and sensible alternatives, and also because *nature* made matters even worse by arranging the cosmic cycles of days and years in such poor and irrational correspondence.

The Pirate King, in Gilbert and Sullivan's leap year tale, begins his explanation of Frederick's calendrical dilemma with the following intonation: "For some ridiculous reason, to which, however, I have no desire to be disloyal . . ." I feel the same way about the day-date problem. For some ridiculous and arbitrary reason, our culture decided to parse days into groups of seven called weeks—a unit with no correspondence whatever to anything cyclical in nature. Because the year runs for 365 days, we end up with fifty-two weeks in a year—and one bothersome extra day left over.

If the year contained a nonfractional number of weeks (I am setting aside the leap year problem for a moment), we would have no day-date problem, for each date of the year would always fall on the same day of the week. But we made matters unduly difficult by running the year through a series of weeks, and then leaving one day over each year—for the day of the week must now shift for each date in each new year. A date that falls on a Tuesday in 1997 must switch to a Wednesday for 1998, Thursday in 1999, and so on. (We will get to those leap years in a moment.)

All this wouldn't matter a damn—except that we care. Don't ask me why, but we take an uncommon interest in the day of the week that our birthday, or some other date of importance to us, occupied in years other than the one now running (which we can easily look up on a calendar). We are especially concerned, in our culture, about the day on the actual date of our birth. Ask almost anyone if they know the weekday of their birth (most people don't), and they will immediately dredge some doggerel out of their infantile consciousness of nursery rhymes: "Monday's child is fair of face, Tuesday's child is full of grace . . ." (Believe me, I am not conjecturing, or making this up. As will soon become clear, I speak from empirical experience and have witnessed the repetition of this scene dozens of times. I was also born on a Wednesday, and "Wednesday's child is full of woe.")

The day-date problem could be solved with relative ease, if the only difficulty lay in this inconvenience of human definition—and the consequent need to add one each year. Anyone could then look up this year's weekday for any date on a calendar, figure out the interval between this year and any other year of interest, divide by seven, and then subtract the remainder from this year's weekday.

But nature now intervenes to impose an additional difficulty based on the fractionality of days in the year, and the consequent need to designate leap years. The year contains fifty-two weeks and one extra day—except for leap years, when the year runs for fifty-two weeks and two extra days. Therefore, to tell the day of the week for any date in a past (or future) year, you must first find the date on this year's calendar. You then need to make two calculations: first, to take *human foibles* into account and correct for the extra day added each year because the year contains fifty-two weeks plus one day; and second, to acknowledge *natural complexities* and make another correction for the extra day added in any leap year (not forgetting, if you are calculating across centuries, the Gregorian rule for omitting leap years at century boundaries not divisible by 400).

The entire procedure therefore becomes pretty complex—and the exercise of figuring out the day of the week for dates in distant years (and doing so quickly enough to keep oneself and others interested) goes by the name of "day-date calculation." The subject has also generated a surprisingly long and learned literature. Some people are prodigious day-date calculators and can instantly (and without error) tell you the day of the week for any date in any year, often ranging widely over centuries and millennia with apparent ease.

As one staple of this literature, some of the most famous, and most proficient, day-date calculators have been mentally retarded or autistic people with general mental skills and accomplishments so limited that no one can figure out how they could possibly develop such an odd, complex, and arcane skill. What could be more marvelous, magical, or even miraculous? Day-date calculation seems hard enough to contemplate for ordinary mortals; how can people with such great limitations possibly manage such a thing? What does their achievement tell us about the nature of human intelligence—not to mention human courage? My final section tells the story of such a person.

PART TWO:
FIVE WEEKS

Poets, extolling the connectedness of all things, have said that the fall of a flower's petal must disturb a distant star. Let us all be thankful that universal integration is not so tight, for we would not even exist in a cosmos of such intricate binding.

Georges Cuvier, the greatest French naturalist of the early nineteenth century, argued that evolution could not occur because all parts of the body are too highly integrated. If one part changed, absolutely every other part would have to alter in a corresponding manner to produce a new but equally elegant configuration for some different mode of life. Since we cannot imagine such comprehensive change of every single part, each to the perfection of a new optimality, organisms cannot evolve.

Half of Cuvier's argument is undeniably sound. If evolution required such comprehensive alteration, such a process might well be impossible. But parts of bodies are largely modular and dissociable to a great extent. Alpha Centauri (not to mention more distant stars) didn't blink the slightest notice when little Susie pulled those petals off the daisy—"He loves me, he loves me not . . ." And even though the foot bone's connected to the ankle bone, evolution can change the number of stripes on a snail's shell without altering the number of teeth on its radula (jaw).

The functions of the brain, and human intelligence in general, also tend to be quite modular and dissociable. No g-factor, or unitary measure of "general intelligence," lurks within the brain, capable of ranking people according to their inherited quantity of a coherent thing, measured by a single number called IQ. (See my critique of this position in my earlier book *The Mismeasure of Man*.) Rather, *intelligence* is a vernacular word that we apply to the large set of relatively independent mental attributes that build, in their entirety, something we call "mind."

The best, and classical, illustration of the relative independence of mental attributes lies in the stunning phenomenon illustrated by people who were once labeled with the stunningly insensitive name *idiot savant*—that is, globally retarded people with a highly precise, separable, and definable skill developed to a degree that would surprise us enough in a person of normal intelligence but that strikes us as simply miraculous in a person otherwise so limited. Some savants can do lightning calculation, multiplying and dividing long strings of numbers instantaneously and with unfailing accuracy—but cannot make change from a dollar or even understand the concept. (Dustin Hoffman played such a character with great sensitivity in *Rain Man*.) Others can draw pictures, accurate to the finest detail, of complex scenes that they have viewed but once and for a fleeting moment—yet cannot read, write, or speak.

These people fascinate us for two very different reasons. We gasp because they are so unusual, and extremes always fascinate us (the biggest, the fiercest, the ugliest, the most brilliant). We need not be ashamed of this quintessentially human propensity. But savants also compel our attention because we feel that they may be able to teach us something important about the nature of normal intelligence—for we often learn most about an average by understanding the reason for an extreme deviation.

We have favored two broad interpretations of these savants (each too simple, and probably both wrong, but still representing a reasonable first pass at formulating the problem). Do these people acquire their extraordinary skill because they discover one thing they can do— and then work so very hard, and so assiduously, at developing it? In this case, any of us could probably master the savant's skill, but we would never choose to devote so much time to one mental operation. (In this alternative, the savant's brain does not differ from ours in the module devoted to his hypertrophied skill—and the phenomenon teaches us something about the nature of dedication.)

Or do these people develop their skill because deficiencies in one part of the brain's structure may be balanced by unusual development in another part? In this case, most of us could not learn the savant's skill even if we chose a path of single-minded devotion to such an activity. (In this alternative, the savant's brain may differ from ours in the module regulating his special skill—and the study of this phenomenon may teach us something important about the physical nature of mentality.)

In any case, day-date calculation represents one of the most famous, and most frequent, of so-called "splinter skills" manifested by many savants. The subject has generated a great deal of study, well summarized in two recent books (Steven B. Smith, *The Great Mental Calculators,* Columbia University Press, 1983; and Darold A. Treffert, *Extraordinary People: Understanding "Idiot Savants,"* Harper and Row, 1989). One obvious question has dominated the literature about mentally retarded and autistic day-date calculators: How do they do it?

The most obvious approach—simply asking a savant how he performs his day-date calculations—does not work. Few of us can give any decent explanation of how we accomplish the things we do best, for our truly unusual accomplishments seem automatic to us. (Sports heroes are famously unable to describe their extraordinary skills—"Well, um, er, I just keep my eye on the ball and . . ." Intellectuals do no better in elucidating their literary or mathematical accomplishments—"Well, um, I had a dream, and I saw these six snakes, and . . .") Savants, if they speak at all, will tend to say "I just do it"—and most of us could describe our special skills no better.

The literature has considered two basic modes—and results are typically inconclusive in illustrating the usual variety of reasons for human achievements. That is, some savants do it one way, others the other way, yet others in combination, and still others in a manner as yet undetermined. First, a savant might have extraordinary, even truly eidetic, skills in memorization. A day-date calculator might then simply memorize the calendars for a certain number of years and read the right day of the week for any date in any year directly out of memory. Second, a savant might develop an algorithm or rule of calculation, and then apply the rule so often, and with such concentration and dedication, that his calculation becomes extremely rapid and "second nature." At some point, the procedure may start to feel automatic.

Some savant day-date calculators do use memory alone—and this method can be spotted because practitioners tend to memorize only a limited number of years. A savant who can do day-date calculation from, say, 1980 to 2020—but has no clue about dates in earlier or later years—has probably memorized forty years' worth of calendars (as researchers might be able to ascertain by checking a subject's bookshelf or asking if he owns a perpetual calendar for a limited number of years).

But many savant day-date calculators, including the young man described herein, use algorithms of their own invention. Some of these people, including my subject, can calculate effortlessly, and apparently instantaneously, sometimes across thousands of years, past or future, and with no apparent difference in the time needed to calculate a date two years or two hundred years from the present. The statement that some savants use algorithms still leaves two mysteries and complexities unaddressed—and these also figure prominently in literature on the subject. First, day-date calculation, as I showed in the last section, is a two-step process. You need, first, to know the day of the week for some reference year—usually the current year as given on a calendar. Then you can apply your algorithm to calculate the difference between your reference year and the year in question. Thus, no matter how good your algorithm, you still need to put some basic reference into memory. (Of course, you could begin any application by looking up the day of the week for this year on a portable calendar—but no self-respecting day-date calculator would use such a crutch.)

Second, and of most potential interest for insight into human mentality in general, the best algorithmic calculators, including my subject, do their reckoning far too quickly to be using their algorithm in an explicit manner. As a striking example, a graduate student studying George and Charles, the famous mathematical twins (and prodigious day-date calculators) so brilliantly and poignantly described by Oliver Sacks (in a chapter in *The Man Who Mistook His Wife for a Hat*), decided to try to equal their skills in day-date calculation by applying their method with the same single-mindedness manifested by many savants. He found that he could do the calculation, but he could not come close to their speed for a long time. Finally, and in a manner that he could never describe accurately, the technique just "clicked" and started to feel automatic. The student could then match the twins. Darold Treffert's book quotes a report by Dr. Bernard Rimland on this experiment:

> Langdon practiced night and day, trying to develop a high degree of proficiency. . . . But despite an enormous amount of practice, he could not match the speed of the twins for quite a long time. Then suddenly, he discovered he could match their speed. Quite to Langdon's surprise, his brain had somehow automated the complex calculations; it had absorbed the table to be memorized so efficiently that now calendar calculating was second nature to him; he no longer had to consciously go through the various operations.

The young man I know, probably one of the best day-date calculators in the nation by now, is autistic and severely limited in cognition. His language skills are good, but his comprehension of intentionality and emotional causality is a virtual blank. He understands basic physical causality, and knows that a dropped object will fall to the ground, or a thrown ball hit the wall, but he cannot read human motivation or the "internal" reasons behind human actions. He cannot understand the simplest story in a book or movie. He can play a game in the sense of learning to follow the rules mechanically, but he has no idea why people engage in such activities and has never begun to grasp such concepts as scoring, winning, and losing.

Humans are storytelling creatures preeminently. We organize the world as a set of tales. How, then, can a person make any sense of his confusing environment if he cannot comprehend stories or surmise human intentions? In all the annals of human heroics, I find no theme more ennobling than the compensations that people struggle to discover and implement when life's misfortunes have deprived them of basic attributes of our common nature.

We tend to understand how the physically handicapped cope, but we rarely consider the similar struggles of the mentally handicapped. We must all order the "buzzing and blooming" confusion of the external world—and if we can't understand stories, we have to find some other way. This young man has struggled all his life to find regularities that might anchor and make sense of the surrounding cacophony. Many of his efforts have been dead ends and wild goose chases.

Since he reads faces so poorly, he struggled for years to find an additional clue in the pitch or loudness of voices. Does high mean happy? Does loud mean angry? He would play the same record at different speeds, converting Paul Robeson at 33 rpm to the sound of a woman's voice at 78 rpm—always hoping (or so I inferred) to induce some rule, some guide to action. He has never found it, though he still tries. When he was quite young, he developed some mathematical skills, and he put them to immediate use. He would time all his 33 rpm records, trying to find some rule that would correlate the type of music with the length of the recording. He got nowhere and eventually gave up.

Finally, he found his workable key—chronology. If you cannot understand stories, what might work next best as a general organizer? The linear sequence of time! You may not know why, or how, or whether, or what, but at least you can order all the items in a temporal series without worrying about their causal connections—this came before that, that before the other, the other before this-thing-here. He had triumphed. This young man can tell you something that happened on every individual day for the last twenty years of his life. Since he does not judge importance as we do, the event that he remembers often seems trivial to us, so we do not recall and therefore cannot verify his accuracy— "On that day, Michael Ianuzzi said 'Wow.' " But when we can check, he is never wrong—"On July 4, 1981, we saw fireworks on the Charles River."

I think I know why he first got interested in day-date calculation. Temporal sequencing had become the touchstone for his ordering of life. And what could be more riveting—and perhaps crucially important in some hidden way—than this interesting change in the weekdays of dates from year to year? There must be some rule behind all this. What could it be? So he struggled and found out. I watched his skills increase, but I never knew how he did the calculation.

If you are going to distinguish yourself by developing a narrow "splinter skill," I cannot imagine a more wonderfully useful choice than day-date calculation. Most people take an interest in knowing the day of the week on the date of their birth. But this information is not easy to come by. You can't look it up in the encyclopedia, and you can't find it on an ordinary calendar. Unless your mother remembered and told you the day, you probably don't know. To find out, you have to be able to perform day-date calculation—and most people can't.

This young man therefore becomes a priceless resource. I have seen him work a room like the best politician. He starts at one end and asks everyone the same question: "What day were you born on, and in what year?" His respondent says, "September 10, 1941," or whatever, and the young man replies without a second's hesitation, and in a special cadence well known to his friends and acquaintances—"A Wednesday." He is never wrong. A half hour later, I see him at the other end of the room. He has made the full circuit with all the aplomb of a diplomat—but with much more genuine interest generated. The feedback is also very gratifying for him—for people want to know and are genuinely grateful. They find his skill inscrutable and amazing—and they tell him so. A little stroking always goes a long way, especially for a man who has tried so hard to comprehend the confusion surrounding him, and has so often failed.

I always understood what this awesome skill in day-date calculation meant to him, but I yearned to find out how he did it—and he could never tell me. I figured out a few bits and pieces. I knew that he worked algorithmically, using this year's calendar (which he knows cold and apparently eidetically) as a reference and starting point. He knows the Gregorian rules for leap years and can therefore extend his calculations instantaneously across centuries and millennia. But what algorithm did he use?

He recognized both components of the general problem—algorithmic day-date calculators must, after all. He knew that the ordinary year contains fifty-two weeks and a day, and that days of the week therefore move forward by one for the same date in subsequent years—this year's Tuesday for any given date becoming next year's Wednesday. He also knew that an additional correction has to be made for leap years. But how did he put these two corrections together? What rule had he devised? I was stymied.

I then spoke to an English TV producer who had made a program on savants. He said to me: "Ask him if there is anything special about the number 28. All savant calculators that I have ever met have discovered this rule." But I didn't know the rule, so I asked him, "What's special about 28?" "Didn't you know?" he replied. "The calendar has a twenty-eight year repeat cycle. This year's calendar is exactly the same as the one for twenty-eight years ago."

Immediately I realized why this must be so—and I figured it out as any ordinary scientist with a modicum of basic mathematics would do. Of course. Two different cycles are operating simultaneously to cause the day-date shifts. First, a seven-year cycle based on the addition of a day each year—so that after seven years (disregarding leap years) the calendar comes back to where it began, and July 10 on a Wednesday becomes July 10 on a Wednesday again. Second, a four-year cycle based on adding an extra leap-year day every four years. So I dredged up an old calculational rule from my schooldays: If two cycles operate together, the multiple of their periods gives you the overall repeat time. Seven times four is twenty-eight. Thus, the calendar must work by a twenty-eight year repeat cycle—and this cycle becomes an obvious key for simplifying day-date calculations. You know the calendar for the current year already. The same calendar works for twenty-eight years ago. 1998 is the same as 1970. You already know that dates for 1999 will move one day of the week forward— and 1971 is the same as 1999. And so it goes.

I had figured this out with some elementary arithmetic, but my autistic friend could not work this way. I was very eager to learn if he knew about the rule of 28. If so, would I finally grasp the key to his algorithm? Would I finally understand how he performed his uncanny lightning calculation? So I asked him: "Is there anything special about the number 28 when you figure out the day of the week for dates in different years?" And he gave me the most beautiful answer that I have ever heard—although I didn't understand a bit of it at first. He said: "Yes . . . five weeks."

I was completely dumbfounded. Obviously, he had misunderstood, and his response had made no sense at all. So I asked again: "Is there anything special about the number 28 when you figure out the day of the week for dates in different years?" And he replied without hesitation: "Yes . . . five weeks."

I understood in a flash several hours later, and his solution was so beautiful that I started to cry. He could not use, or even understand, my arithmetical rule about multiplying the periods of two different cycles together. He could only work by counting concrete days, one after the other. He had figured out the following principle by thinking concretely in the only manner available to him: A year contains fifty-two weeks and some extra days—one extra day in an ordinary year, two extra days in a leap year. When the total number of extra days becomes evenly divisible by seven, then the calendar for that year is the same as the calendar I already know for this year. (The same argument works by subtracted days for past years, or by added days for future years.) If I can figure out a minimum span of years for which the number of added days is always exactly the same, and always exactly divisible by seven, then the calendar must repeat and I will have my rule.

So he began to count the number of added days concretely, one by one, year by year. Every span of years up to 28 couldn't work because the number of leap years varies. Thus, for example, a thirteen-year period may have four leap years (1960–1972) or three leap years (1961–1973). But when you reach 28 years—and never before—everything works out just right. Every 28-year span, whenever you start and wherever you finish, contains exactly seven leap years. (I am disregarding the Gregorian rule for omitting leap years at most century boundaries. As all day-date calculators know, this situation requires a special correction—and you must keep track of it separately.) Every 28-year span also includes exactly 28 extra days, arising from the rule that every year adds one day. Thus, every interval of 28 years adds exactly 35 days, no more, no less—one for each of the 28 years, plus seven additional days for the invariable number of leap years. Since 35 is exactly divisible by 7, the calendar must repeat every 28 years.

I now finally understood how this consummate day-date calculator worked. He had added extra days concretely, the only mental method available to him. He could not use my mindless, memorized schoolboy rule—I still don't really know why it works—of multiplying the periods of coincident cycles together. He had added up extra days laboriously until he came to 28 years—the first span that always adds exactly the same total number of extra days, with the sum of extra days exactly divisible by seven. Every 28 years includes 35 extra days, and 35 extra days makes five weeks. You see, he *had* given me the right answer to my question—but I had not understood him at first. I had asked: "Is there anything special about the number 28 when you figure out the day of the week for dates in different years?" and he had answered: "Yes . . . five weeks."

May we all make such excellent use of our special skills, whatever and however limited they may be, as we pursue the most noble of all our mental activities in trying to make sense of this wonderful world, and the small part we must play in the history of life. Actually, I didn't quote his beautiful answer fully. He said to me: "Yes, Daddy, five weeks." His name is Jesse. He is my firstborn son, and I am very proud of him.

ILLUSTRATION CREDITS

Grateful acknowledgment is made for permission to reproduce the following:

Frontispiece: Detail from *The Last Judgement* (1536–1541) by Michelangelo Buonarroti. Sistine Chapel, Vatican Place, Vatican State. Courtesy of Alinari/Art Resource, New York.

Page 34: The jaws of hell fastened by an angel, from the *Psalter of Henry of Bloise, Bishop of Winchester*, twelfth century. British Library, London, Great Britain. Courtesy of Bridgeman/Art Resource, New York.

Page 42: *Sinners in Hell, Last Judgement,* anonymous. Relief. Mainz, Germany. Courtesy of Foto Marburg/Art Resource, New York.

Page 51: *Apocalypse of Saint John: Babylon Falls on the Demons* (1363–1400) by Nicolas Bataille. Tapestry. Musée des Tapisseries, Angers, France. Courtesy of Giraudon/Art Resource, New York.

Page 63: Detail from *The Last Judgement* by Fra Angelico (1387–1455). Galleria d'Arte Moderna, Florence, Italy. Courtesy of Alinari/Art Resource, New York.

Page 70: *The Last Judgment* by an anonymous Bologna artist, fourteenth century. Pinacoteca Nazionale, Bologna, Italy. Courtesy of Scala/Art Resource, New York.

Page 79: Detail from *The Last Judgement* (1443) by Rogier van der Weyden. Altarpiece, center panel. Hôtel-Dieu, Beaune, France. Courtesy of Giraudon/Art Resource, New York.

Page 90: *The Opening of the Fifth and Sixth Seals, the Distribution of White Garments Among the Martyrs and the Fall of Stars* (1498) by Albrecht

INDEX

Italicized page numbers indicate illustrations

INDEX

A SELECTED LIST OF NON-FICTION
AVAILABLE IN VINTAGE

☐	SLOW RECKONING	Tom Athanasiou	£7.99
☐	THIS TIME	Anthony Barnett	£6.99
☐	GUNS, GERMS AND STEEL	Jared Diamond	£8.99
☐	BURY ME STANDING	Isabel Fonseca	£7.99
☐	THE KILLING OF THE COUNTRYSIDE	Graham Harvey	£7.99
☐	WESTMINSTER WOMEN	Linda McDougall	£8.99
☐	HOW WE LIVE	Sherwin Nuland	£7.99
☐	HIDDEN AGENDAS	John Pilger	£8.99
☐	COVERING ISLAM	Edward W. Said	£7.99
☐	THE END OF TIME	Damian Thompson	£6.99
☐	PROMISCUITIES	Naomi Wolf	£7.99

- All Vintage books are available through mail order or from your local bookshop.

- Please send cheque/eurocheque/postal order (sterling only), Access, Visa, Mastercard, Diners Card, Switch or Amex:

☐☐☐☐☐☐☐☐☐☐☐☐☐☐☐

Expiry Date:_____Signature:_____

Please allow 75 pence per book for post and packing U.K.
Overseas customers please allow £1.00 per copy for post and packing.

ALL ORDERS TO:

Vintage Books, Books by Post, TBS Limited, The Book Service,
Colchester Road, Frating Green, Colchester, Essex CO7 7DW

NAME:_____

ADDRESS:_____

_____

_____

Please allow 28 days for delivery. Please tick box if you do not
wish to receive any additional information   ☐

Prices and availability subject to change without notice.